Mathematics in Architecture, Art, Nature, and Beyond

Problem Solving in Mathematics and Beyond

Print ISSN: 2591-7234
Online ISSN: 2591-7242

Series Editor: Dr. Alfred S. Posamentier
Distinguished Lecturer
New York City College of Technology - City University of New York

---

There are countless applications that would be considered problem solving in mathematics and beyond. One could even argue that most of mathematics in one way or another involves solving problems. However, this series is intended to be of interest to the general audience with the sole purpose of demonstrating the power and beauty of mathematics through clever problem-solving experiences.

Each of the books will be aimed at the general audience, which implies that the writing level will be such that it will not engulfed in technical language — rather the language will be simple everyday language so that the focus can remain on the content and not be distracted by unnecessarily sophiscated language. Again, the primary purpose of this series is to approach the topic of mathematics problem-solving in a most appealing and attractive way in order to win more of the general public to appreciate his most important subject rather than to fear it. At the same time we expect that professionals in the scientific community will also find these books attractive, as they will provide many entertaining surprises for the unsuspecting reader.

Published

Vol. 37 *Mathematics in Architecture, Art, Nature, and Beyond*
by Alfred S Posamentier and Günter J Maresch

Vol. 36 *Geometric Gems: An Appreciation for Geometric Curiosities Volume III: The Wonders of Circles*
by Alfred S Posamentier and Robert Geretschläger

Vol. 35 *Geometric Gems: An Appreciation for Geometric Curiosities Volume II: The Wonders of Quadrilaterals*
by Alfred S Posamentier and Robert Geretschläger

Vol. 34 *A Journey Through the Wonders of Plane Geometry*
by Alfred S Posamentier and Hans Humenberger

For the complete list of volumes in this series, please visit www.worldscientific.com/series/psmb

Problem Solving in Mathematics and Beyond

Volume 37

Mathematics in Architecture, Art, Nature, and Beyond

Alfred S. Posamentier
The City University of New York, USA

Günter J. Maresch
University of Salzburg, Austria

World Scientific

NEW JERSEY • LONDON • SINGAPORE • BEIJING • SHANGHAI • TAIPEI • CHENNAI

Published by

World Scientific Publishing Co. Pte. Ltd.
5 Toh Tuck Link, Singapore 596224
USA office: 27 Warren Street, Suite 401-402, Hackensack, NJ 07601
UK office: 57 Shelton Street, Covent Garden, London WC2H 9HE

Library of Congress Cataloging-in-Publication Data
Names: Posamentier, Alfred S, author. | Maresch, Günter J., author.
Title: Mathematics in architecture, art, nature, and beyond / Alfred S Posamentier, Günter J Maresch.
Description: New Jersey : World Scientific, [2025] | Series: Problem solving in mathematics and beyond, 2591-7234 ; vol. 37 | Includes bibliographical references and index.
Identifiers: LCCN 2024056914 | ISBN 9789811296987 (hardcover) | ISBN 9789811296994 (ebook for institutions) | ISBN 9789811297007 (ebook for indivduals)
Subjects: LCSH: Mathematics--Study and teaching (Secondary) | Mathematics in nature. | Art--Mathematics. | Architecture--Mathematics.
Classification: LCC QA11.2 .P639 2025 | DDC 510.71/2--dc23/eng/20250211
LC record available at https://lccn.loc.gov/2024056914

British Library Cataloguing-in-Publication Data
A catalogue record for this book is available from the British Library.

First published 2025 (hardcover)
Reprinted 2026 (in paperback edition)
ISBN 978-981-98-3048-0 (pbk)

For any available supplementary material, please visit
https://www.worldscientific.com/worldscibooks/10.1142/13954#t=suppl

Desk Editors: Murali Appadurai/Rosie Williamson

Typeset by Stallion Press
Email: enquiries@stallionpress.com

Preface

There are few opportunities to show the amazing applications of mathematics outside the classroom. To motivate students, some teachers allude to the usefulness of mathematics in everyday life. However, everyday life varies amongst many students, and so this attempt to motivate pupils by merely mentioning the potential practicality of mathematics—often without applications—is sometimes ineffective. The mathematics teacher must cover the entire prescribed curriculum, which leaves little or no time to introduce students to the wonders of the subject. This book shows how mathematics manifests in a wide variety of areas, manifestations that most people are unaware of. One can even revel in the evolution of our number system and how that has enabled us to define the beauty in mathematics as well as the beauty in art, architecture, and beyond.

The current number system today, often referred to as the Hindu-Arabic number system, was introduced to the European world by Leonardo of Pisa, better known as Fibonacci. In his book *Liber abaci*, Fibonacci presents the numerals 1, 2, 3, 4, 5, 6, 7, 8, 9, and the 0, which he learned as a youngster working with mathematicians in the Arab world. This was the first appearance of these numerals in the European culture, and yet it took about another 50 years before the Hindu-Arabic system became the standard. *Liber abaci* is a book of mathematical problems, and a problem in the twelfth chapter concerning the regeneration of rabbits publicized the now famous Fibonacci numbers. These ubiquitous numbers can be seen in countless aspects of our environment.

In the first chapter, we introduce the Fibonacci numbers and their amazing relationships. This is followed in the next chapter by applications of these numbers in our general environment. The ratio of any two consecutive Fibonacci numbers approaches the Golden Ratio as the numbers get larger and larger. Many believe that the Golden Ratio helps define generalized beauty; whether in architecture or any plane figures, the Golden Ratio is universally aesthetically pleasing. The third chapter illustrates how the Golden Ratio manifests itself in a wide variety of visual aspects, such as architecture and artwork. The Golden Ratio is also often seen as the Golden Rectangle, which is a rectangle the ratio of whose dimensions form the Golden Ratio. This leads to a discussion of geometry in architecture, nature, including plants, geology, and man-made designs such as logos in fourth, fifth and sixth chapters. The final chapter concludes our enjoyable journey through the world's mathematical wonders with conic sections and their application in radar dishes, headlight reflectors, whispering halls, and much more.

We hope that exposure to aspects of mathematics that are usually bypassed in the classroom will foster a new love for a subject that is un- and underappreciated despite its dominance in the school curriculum. The goal of this book is to provide the general readership with an opportunity to truly appreciate the power and the beauty of mathematics.

About the Authors

Alfred S. Posamentier is currently Distinguished Lecturer at New York City College of Technology of the City University of New York. Previously, he was the Executive Director for Internationalization and Sponsored Programs at Long Island University, New York. This was preceded by a five-year period, where he was Dean of the School of Education and tenured Professor of Mathematics Education at Mercy University, New York. He is now also Professor Emeritus of Mathematics Education at The City College of the City University of New York, and former Dean of the School of Education, where he was tenured for 40 years. He is the author and co-author of more than 80 mathematics books for teachers, secondary and elementary school students, and the general readership. Dr. Posamentier is also a frequent commentator in newspapers and journals on topics relating to education. After completing his B.A. degree in mathematics at Hunter College of the City University of New York in 1964, he took a position as a teacher of mathematics at Theodore Roosevelt High School (Bronx, NY), where he focused his attention on improving the students' problem-solving skills and at the same time enriching their instruction far beyond what the traditional textbooks offered. During his six-year tenure there, he also developed the school's first mathematics teams (both at the junior and senior level). He is still involved in working with mathematics teachers and supervisors, nationally and internationally, to help them maximize their effectiveness. During this time, he earned an M.A. degree at the City College of the City University of New York in 1966. Immediately upon joining the faculty of the City College in 1970, he began to develop in-service courses for secondary school mathematics teachers, including such special areas as recreational mathematics and problem-solving in mathematics. As Dean of the City College School of Education for 10 years, his scope of interest in educational issues covered the full gamut. During his tenure as dean, he took the School of Education from the bottom of the New York State rankings to the top with a perfect NCATE accreditation assessment in 2009. He achieved the same success in 2014 at Mercy University, which received both NCATE and CAEP accreditation during his leadership as Dean of the School of Education.

In 1973, Dr. Posamentier received his Ph.D. from Fordham University (New York) in mathematics education and has since extended his reputation in mathematics education to Europe. He has been visiting professor at several European universities in Austria, England, Germany, Czech Republic, and Poland, while at the University of Vienna he was Fulbright Professor (1990). In 1989 he was awarded an Honorary Fellow at the South Bank University (London, England). In recognition of his outstanding teaching, the City College Alumni Association named him Educator of the Year in 1994, and in 2009. New York City had the day, May 1, 1994, named in his honor by the President of the New York City Council. In 1994, he was also awarded the Grand Medal of Honor from the Republic of Austria, and in 1999, upon approval of Parliament, the President of the Republic of Austria awarded him the title of University Professor of Austria. In 2003 he was awarded the title of Ehrenbürger (Honorary Fellow) of the Vienna University of Technology, and in 2004 was awarded the Austrian Cross of Honor for Arts and Science, First Class from the President of the Republic of Austria. In 2005 he was inducted into the Hunter College Alumni Hall of Fame, and in 2006 he was awarded the prestigious Townsend Harris Medal by the City College Alumni Association. He was inducted into the New York State Mathematics Educator's Hall of Fame in 2009, in 2010 he was awarded the coveted Christian-Peter-Beuth Prize in Berlin, and in 2017 he received the Summa Cum Laude nemine discrepante Award from Fundacion Sebastian, A.C. in Mexico City. He has taken on numerous important leadership positions in mathematics education locally. He was a member of the New York State Education Commissioner's Blue-Ribbon Panel on the Math-A Regents Exams in 2003, and in 2014–15 he was on the Commissioner's Mathematics Standards Committee, which redefined the Standards for New York State, and he also served on the New York City schools' Chancellor's Math Advisory Panel. Dr. Posamentier is a leading commentator on educational issues and continues his long-time passion of seeking ways to make mathematics interesting to both teachers, students and the general public — as can be seen from some of his more recent books: See https://en.wikipedia.org/wiki/Alfred_S._Posamentier.

Günter J. Maresch is currently Professor of Mathematics Education and Vice-Dean of the Faculty of Digital and Analytical Sciences at the University of Salzburg (Austria). His research topics are spatial ability, spatial thinking training and diagnosis, descriptive geometry, computer-aided design (CAD), new digital media, didactical principles, and curriculum development. Dr. Maresch studied descriptive geometry and mathematics at the Vienna University of Technology (Austria) and after completing his Master's degree (1994), he took a position as a teacher of mathematics, descriptive geometry, and information and communication technology (ICT) at secondary schools in the province of Salzburg (Austria). Dr. Maresch was a lecturer at the University of Salzburg from 1998 to 2014, where he taught courses on spatial ability, descriptive geometry, computer-aided design (CAD), new digital media, and curriculum development. Dr. Maresch completed his Ph.D. studies at the University of Salzburg (Austria) in 2004. His Ph.D. thesis focused on e-learning and computer-aided design as well as the development of a didactical concept for increasing the quality of geometry instruction. He continues to provide professional programs in many countries, including Austria, Germany, Finland, Netherlands, Denmark, Spain, Hungary, and South Africa. From 2014 to 2016, he was an Assistant Professor for Mathematics Education and Geometry at the University of Salzburg. In 2016, Dr. Maresch was awarded the rank of Associate professor for Mathematics Education at the University of Salzburg.

Since 1998 Dr. Maresch has provided lecturers at more than 300 conferences and teacher in-service training programs in the form of keynotes, lectures, workshops, and seminars, which were focused on geometry (spatial ability, curriculum, new media, competence model(s), CAD trainings, etc.), mathematics (educational standards, competence model, didactical principles, etc.), ICT (new digital media, learning platforms, computer-aided design, etc.), principal conferences and school inspector conferences (sustainability of teacher in-service trainings, etc.), and supervising teacher internship (centralized final exams, prescientific research paper, pedagogical competences, etc.). Dr. Maresch completed studies to earn a teaching certificate for ICT at the local school board of Salzburg (1998). He further qualified for the Intel Master Teacher Certificate (2001) and completed the Leadership Academy Program (2008–2009) offered by the Austrian Ministry of Education. From 1998 to 2007, Dr. Maresch was the Teachers' Mentor for teachers of descriptive geometry, mathematics, and ICT in the province of Salzburg. He also organized more than 50 conferences and teacher in-service training programs during this time.

In 2012, Dr. Maresch was named "Educator of the Year" at the worldwide competition of Bentley Systems, and in 2020, he was one of the winners of the "Digital Teaching Award" of the University of Salzburg. From 2007 to 2012, Dr. Maresch was a member of the local education management board and the head of the institute for lifelong learning for upper secondary schools at the Salzburg University College of Teacher Education and during this time he organized more than 700 teacher in-service training programs for teachers and headmasters of Austrian lower and upper secondary schools per year. Since 2002, he has been a member of the curriculum board of the Austrian Ministry of Education—responsible for the development of the curricula for the secondary schools.

In summary, he has published more than 100 research papers and 20 books for students, researchers and teachers of mathematics, and books on geometry for the general readership.

Dr. Maresch is the inventor and chief developer of the international online spatial thinking training and diagnosis platform RIF (https://rif4you.eu/en). More than 3 million tasks have been completed on the free platform by students from 49 countries around the world.

Acknowledgements

We would like to thank the following people for their valuable contributions to this book.

Markus Bilo is a graduate student at the University of Salzburg, Austria and is to be thanked for his contributions to Chapter 4.

Christoph Vierthaler is a teacher at the Holztechnikum Kuchl (Austria) and is in a master's degree in mathematics program at the University of Salzburg, Austria. He is to be thanked for his contributions to Chapter 6.

Hanna Wegscheider is a research assistant in the research group for Didactics of Mathematics (Department of Mathematics) at the University of Salzburg, Austria, and is to be thanked for her contributions to Chapter 5.

Contents

Chapter 1

Introduction to the Fibonacci Numbers

To begin our journey through the visual aspects of mathematics it would be appropriate to introduce the most famous numbers in mathematics. These numbers encroach on all aspects both visual and conceptual. However, to fully appreciate these numbers, we will start with a bit of history and then explore the numbers before we embark upon their visual aspects.

At the beginning of the 13th century, Europe began to wake from the long sleep of the Middle Ages and perceive faint glimmers of the coming Renaissance. By the end of the century, Marco Polo had journeyed the Great Silk Road to reach China, Giotto di Bondone (Giotto) had changed the course of painting and freed it from Byzantine conventions, and the mathematician Leonardo Pisano, or Leonardo of Pisa, best known as Fibonacci,[1] changed forever Western methods of calculation and at the same time indirectly introduced a sequence of numbers that to this day fascinates mathematicians. Many keep coming up with further appearances of these numbers, and the Fibonacci Association, founded in 1963, is a tribute to the enduring contributions of the master.

Leonardo Pisano was born to Guilielmo Bonacci and his wife in the port city of Pisa, Italy, in 1175. Pisa had played a powerful role in commerce since Roman times. In 1192, Guilielmo Bonacci became a public clerk in the

[1]It is unclear who first used the name Fibonacci; however, it seems to be attributed to Giovanni Gabriello Grimaldi (1757–1837) at around 1790, or to Pietro Cossali (1748–1815).

Figure 1-1 Leonardo of Pisa (Fibonacci)

customs house for the Republic of Pisa, and was stationed in the Pisan colony of Bugia (today Bejaia, Algeria) on the Barbary Coast of Africa. Shortly after his arrival he brought his son, Leonardo, to join him so that the boy could learn the skill of calculating and become a merchant. It was in Bugia that Fibonacci first became acquainted with the "nine Indian figures," as he called the Hindu numerals, and "the sign 0 which the Arabs call zephyr," according to his most famous book, *Liber abaci*. During his time away from Pisa, he received instruction from a Muslim teacher who introduced him to a book on algebra entitled *Hisâb al-jabr w'al-muqabâlah*[2] by the Persian mathematician al-Khowarizmi (ca. 780 – ca. 850), which also influenced him. *Liber abaci*,[3] which Fibonacci wrote in 1202 and revised in 1228, is largely a collection of algebraic problems and those "real world" problems that require more abstract mathematics. Fibonacci wanted to spread these

[2]The word "algebra" comes from the title of this book.

[3]The title means "book on calculation." It is not about the abacus.

newfound techniques and the newly discovered Hindu-Arabic numerals to his countrymen. During these times, there was no printing press, so books had to be written by the hand of scribes. Therefore, we are fortunate to still have copies of *Liber abaci*.[4]

The Book, *Liber abaci*

Although Fibonacci wrote several books, we will focus on *Liber abaci*. This extensive volume is full of very interesting problems. Based on the arithmetic and algebra, which Fibonacci had accumulated during his travels, it was widely copied and imitated, and as noted, introduced into Europe the Hindu-Arabic place-valued decimal system, along with the use of Arabic numerals. The book was widely used for the better part of the next two centuries—a bestseller!

Fibonacci begins the book *Liber abaci* with the following:

> The nine Indian figures are: **9 8 7 6 5 4 3 2 1**.[5]
>
> With these nine figures, and with the sign **0**, which the Arabs call zephyr, any number whatsoever is written, as demonstrated below. A number is a sum of units, and through the addition of them the number increases by steps without end. First, one composes those numbers, which are from one to ten. Second, from the tens are made those numbers, which are from ten up to one hundred. Third, from the hundreds are made those numbers, which are from one hundred up to one thousand. ... and thus, by an unending sequence of steps, any number whatsoever is constructed by joining the preceding numbers. The first place in the writing of the numbers is at the right. The second follows the first to the left.

[4]Baldassare Boncompagni translated *Liber abaci* from Latin into Italian in 1857, the first translation into a modern language. An English language version of *Liber abaci* was recently published by Laurence E. Sigler (New York: Springer-Verlag, 2002).

[5]It is assumed that Fibonacci wrote the numerals in order from right to left, since he took them from the Arabs who write in this direction.

Despite their relative facility, these numerals were not widely accepted by merchants who were suspicious of those who knew how to use them. They were simply afraid of being cheated. It took quite a while for these numbers to be accepted throughout the European world. Interestingly, *Liber abaci* also contains simultaneous linear equations. Many of the problems that Fibonacci considers, however, were similar to those appearing in Arab sources. This does not detract from the value of the book, since it is the solutions to these problems that make the major contribution to our development of mathematics. As a matter of fact, a number of mathematical terms—common in today's usage—were first introduced in *Liber abaci*. Fibonacci referred to "factus ex multiplicatione"[6] and from this first sighting of the word, we speak of the "factors of a number" or the "factors of a multiplication." Another example of words seemingly introduced into the current mathematics vocabulary by this famous book are "numerator" and "denominator."

Some of the classical problems introduced by Fibonacci, and which are considered recreational mathematics today, first appeared in the Western world in *Liber abaci*. Yet the technique for solution was always the chief concern for introducing a problem. This book is of interest to us, not only because it was the first publication in Western culture to use the Hindu numerals to replace the clumsy Roman numerals, or because Fibonacci was the first to use a horizontal fraction bar, but because it casually includes a recreational mathematics problem in chapter 12 that has made Fibonacci famous for posterity. This is the problem on the regeneration of rabbits.

The Rabbit Problem

The rabbit problem as it appears in *Liber abaci* is shown in Figure 1-2. The translation of this most important page is provided in Figure 1-3, while Figure 1-3 shows how the problem was stated (with the marginal notes included).

---

[6]David Eugene Smith, *History of Mathematics*, Vol. 2, (New York: Dover, 1958), p. 105.

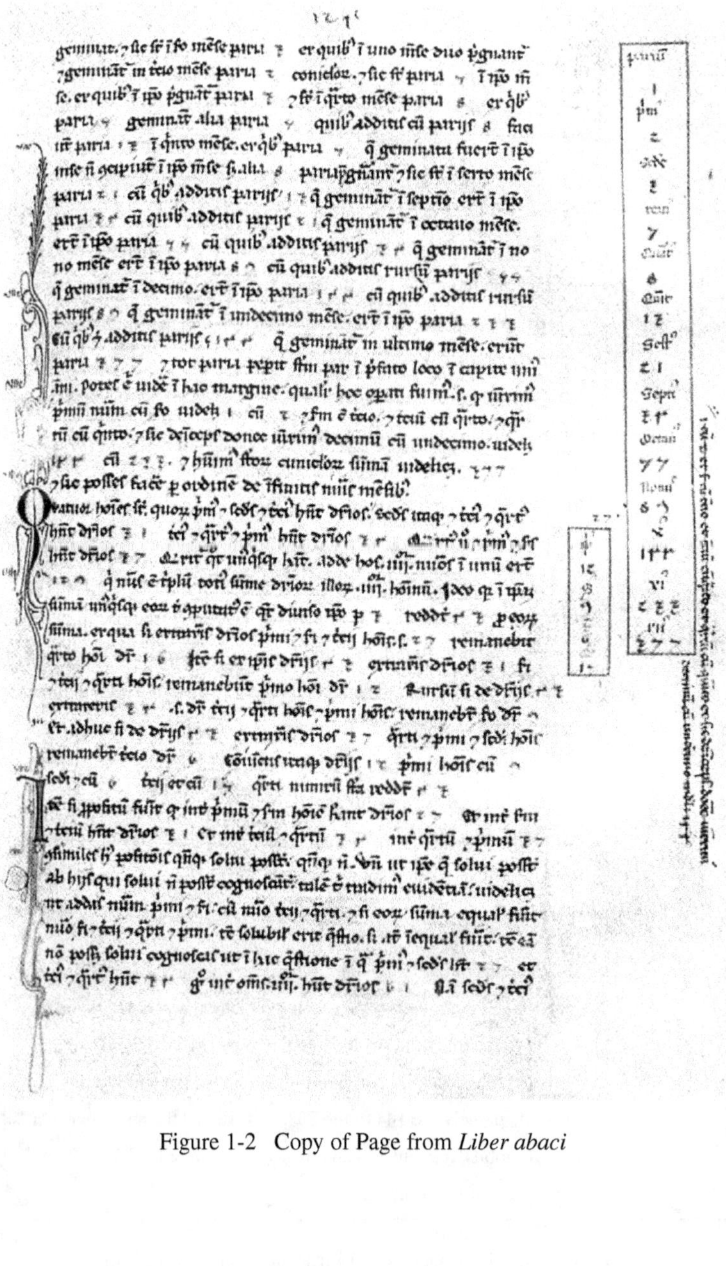

Figure 1-2 Copy of Page from *Liber abaci*

Beginning 1 First 2 Second 3 Third 5 Fourth 8 Fifth 13 Sixth 21 Seventh 34 Eighth 55 Ninth 89 Tenth 144 Eleventh 233 Twelfth 377	"A certain man had one pair of rabbits together in a certain enclosed place, and one wishes to know how many are created from the pair in one year when it is the nature of them in a single month to bear another pair, and in the second month those born to bear also. Because the above written pair in the first month bore, you will double it; there will be two pairs in one month. One of these, namely the first, bears in the second month, and thus there are in the second month 3 pairs; of these in one month two are pregnant and in the third month 2 pairs of rabbits are born and thus there are 5 pairs in the month; in this month 3 pairs are pregnant and in the fourth month there are 8 pairs, of which 5 pairs bear another 5 pairs; these are added to the 8 pairs making 13 pairs in the fifth month; these 5 pairs that are born in this month do not mate in this month, but another 8 pairs are pregnant, and thus there are in the sixth month 21 pairs; to these are added the 13 pairs that are born in the seventh month; there will be 34 pairs in this month; to this are added the 21 pairs that are born in the eighth month; there will be 55 pairs in this month; to these are added the 34 pairs that are born in the ninth month; there will be 89 pairs in this month; to these are added again the 55 pairs that are both in the tenth month; there will be 144 pairs in this month; to these are added again the 89 pairs that are born in the eleventh month; there will be 233 pairs in this month. To these are still added the 144 pairs that are born in the last month; there will be 377 pairs and this many pairs are produced from the above-written pair in the mentioned place at the end of one year. You can indeed see in the margin how we operated, namely that we added the first number to the second, namely the 1 to the 2, and the second to the third and the third to the fourth and the fourth to the fifth, and thus one after another until we added the tenth to the eleventh, namely the 144 to the 233, and we had the above-written sum of rabbits, namely 377 and thus you can in order find it for an unending number of months."

Figure 1-3 Fibonacci's Rabbit Regeneration Problem

Month	Pairs	No. of Pairs of Adults (A)	No. of Pairs of Babies (B)	Total Pairs
Jan. 1	A	1	0	1
Feb. 1	A B	1	1	2
Mar. 1	A B A	2	1	3
Apr. 1	A B A A B	3	2	5
May 1	A B A A B A B A	5	3	8
June 1	A B A A B A B A A B A A B	8	5	13
July 1		13	8	21
Aug. 1		21	13	34
Sept. 1		34	21	55
Oct. 1		55	34	89
Nov. 1		89	55	144
Dec. 1		144	89	233
Jan. 1		233	144	377

Figure 1-4

Consider the chart in Figure 1-4, to see how the number of rabbits increases each month. If we assume that a pair of baby rabbits (B) mature in one month to become offspring-producing adults (A), the chart in Figure 1-5 provides a summary of the number of rabbits existing each month. The solution to this problem, which shows the number of rabbits each month essentially generates the famous sequence of numbers

$$\mathbf{1, 1, 2, 3, 5, 8, 13, 21, 34, 55, 89, 144, 233, 377, \ldots,}$$

known today as the *Fibonacci sequence*. At first glance there is nothing spectacular about these numbers beyond the relationship that would allow us to generate additional numbers of the sequence quite easily by simply adding two consecutive numbers to get the next number, in other words,

1

1

1 + 1 = **2**

1 + 2 = **3**

2 + 3 = **5**

3 + 5 = **8**

5 + 8 = **13**

8 + 13 = **21**

13 + 21 = **34**

21 + 34 = **55**

34 + 55 = **89**

55 + 89 = **144**

89 + 144 = **233**

144 + 233 = **377**

233 + 377 = **610**

377 + 610 = **987**

610 + 987 = **1597 ...**

Figure 1-5 The Fibonacci Sequence

every number in the sequence (after the first two) is the sum of the two preceding numbers.

The Fibonacci sequence can be written in a way that its recursive definition becomes clear: each number is the sum of the two preceding numbers, as shown in Figure 1-5. The Fibonacci sequence is the oldest known (recursive) *recurrent* sequence. There is no evidence that Fibonacci knew of this relationship, but it is securely assumed that a man of his talents and insight knew the recursive relationship. It took another 400 years before this relationship appeared in print.

Introducing the Fibonacci Numbers

These numbers were not identified as anything special when Fibonacci wrote *Liber abaci*. As a matter of fact, the famous German mathematician and astronomer Johannes Kepler (1571–1630) mentioned these numbers in

Figure 1-6 François-Édouard-Anatole Lucas
Source: http://edouardlucas.free.fr/ with permission.

a 1611 publication[7] when he said that the ratios "as 5 is to 8, so is 8 to 13, so is 13 to 21 almost." Centuries passed and the numbers still went unnoticed. In the 1830s, the German botanists Karl Friedrich Schimper (1803–1867) and Alexander Braun (1805–1877) noticed that the Fibonacci numbers appeared as the number of spirals of bracts on a pinecone. Throughout the mid-1800s the Fibonacci numbers captured the fascination of mathematicians. They were given their current name ("Fibonacci numbers") by French mathematician François-Édouard-Anatole Lucas (1842–1891), usually referred to as "Edouard Lucas," who later devised his own sequence by following the pattern set by Fibonacci. Instead of starting with 1, 1, 2, 3, . . . , Lucas began his sequence with 1, 3, 4, 7, 11, ... Had he started with 1, 2, 3, 5, 8, . . . , he would have ended up with a somewhat truncated version of the Fibonacci sequence. (We will inspect the Lucas numbers later in this chapter.)

At about this time, the French mathematician Jacques-Philippe-Marie Binet (1786–1856) developed a formula for finding any Fibonacci number given its position in the sequence. That is, with Binet's formula we can

[7] Maxey Brooke, "Fibonacci Numbers and Their History Through 1900," *Fibonacci Quarterly*, 2:2 (April 1964), 149.

find the 118th Fibonacci number without having to list the previous 117 numbers. The *Binet Formula* for the nth Fibonacci number is

$$\frac{1}{\sqrt{5}}\left[\phi^n - \left(-\frac{1}{\phi}\right)^n\right] = \frac{1}{\sqrt{5}}\left[\left(\frac{1+\sqrt{5}}{2}\right)^n - \left(\frac{1-\sqrt{5}}{2}\right)^n\right],$$

where ϕ is the Golden Ratio, which will be presented later in the book. Essentially, this formula will give us any Fibonacci number for any natural number n.

One may ask, what is so special about these numbers? That is what we hope to demonstrate throughout this chapter. Before we explore the vast variety of examples in which we find the Fibonacci numbers, let us first inspect this famous Fibonacci number sequence and some of the remarkable properties it has.

We will use the symbol F_7 to represent the 7th Fibonacci number, and F_n to represent the nth Fibonacci number, or as we say, the general Fibonacci number, that is, any Fibonacci number. Let us look at the first 30 Fibonacci numbers (Figure 1-7). Notice that after we begin with the first two 1s, in every case after them, each number is the sum of the two preceding numbers.

Some Properties of the Fibonacci Numbers

To fully appreciate the unusual aspect of these numbers, we will present some of the countless curious characteristics of the Fibonacci sequence.

Property 1

The sum of any ten consecutive Fibonacci numbers is divisible by 11. We could convince ourselves that this may be true by considering some randomly chosen examples. Take, for example, the sum of the following ten consecutive Fibonacci numbers:

$$13 + 21 + 34 + 55 + 89 + 144 + 233 + 377 + 610 + 987 = 2{,}563,$$

which just happens to be divisible by 11, since $11 \cdot 233 = 2{,}563$.

$F_1 = 1$
$F_2 = 1$
$F_3 = 2$
$F_4 = 3$
$F_5 = 5$
$F_6 = 8$
$F_7 = 13$
$F_8 = 21$
$F_9 = 34$
$F_{10} = 55$
$F_{11} = 89$
$F_{12} = 144$
$F_{13} = 233$
$F_{14} = 377$
$F_{15} = 610$
$F_{16} = 987$
$F_{17} = 1597$
$F_{18} = 2584$
$F_{19} = 4181$
$F_{20} = 6765$
$F_{21} = 10946$
$F_{22} = 17711$
$F_{23} = 28657$
$F_{24} = 46368$
$F_{25} = 75025$
$F_{26} = 121393$
$F_{27} = 196418$
$F_{28} = 317811$
$F_{29} = 514229$
$F_{30} = 832040$

Figure 1-7

Consider another example: The sum of another ten consecutive Fibonacci numbers, say, from F_{21} to F_{30}:

$$10{,}946 + 17{,}711 + 28{,}657 + 46{,}368 + 75{,}025 + 121{,}393 + 196{,}418$$
$$+317{,}811 + 514{,}229 + 832{,}040 = 2{,}160{,}598,$$

which also is a multiple of 11, since $11 \cdot 196{,}418 = 2{,}160{,}598$.

Property 2

Two consecutive Fibonacci numbers do not have any common factors, which means that they are said to be relatively prime.[8] This can be seen by mere inspection, or take any two consecutive Fibonacci numbers and factor them and you will see that they have no common factors. Take a look at Figure 1-8, where we have listed the first 40 Fibonacci numbers and factored those that are not prime. Notice that no two consecutive Fibonacci numbers have any common factors.

Property 3

Consider the Fibonacci numbers in the composite-number[9] positions (with the exception of the 4th Fibonacci number). That is, we will consider the following Fibonacci numbers: 6th, 8th, 9th, 10th, 12th, 14th, 15th, 16th, 18th, 20th, and so on, which are all non-prime Fibonacci numbers. A quick inspection (Figure 1-9) shows that the Fibonacci numbers in the composite-number (i.e., non-prime) positions are also composite numbers.

Property 4

One would conjecture at this point that the analog of this is also true. Namely, that the Fibonacci numbers in prime-number[10] positions are also prime. That is, if we look at the list of the first 30 Fibonacci numbers shown in Figure 1-10, those in the prime positions, namely, the 2nd, 3rd, 5th, 7th, 11th, 13th, 17th, 19th, 23rd, and 29th Fibonacci numbers would then also have to be prime. But this is not the case, as you will notice that the 2nd and the 19th Fibonacci numbers are not primes and, therefore, this analog situation does *not* hold true. One counterexample suffices to draw this conclusion.

---

[8]Two numbers (integers) are relatively prime if they have no common factors other than 1.
[9]A composite number is one that is not prime; that is, it is divisible by numbers other than itself and 1.
[10]A prime number is a number (other than 0 or ±1) that is divisible only by itself and 1.

n	Fn	Factors
1	1	unit
2	1	unit
3	2	prime
4	3	prime
5	5	prime
6	8	2^3
7	13	prime
8	21	$3 \cdot 7$
9	34	$2 \cdot 17$
10	55	$5 \cdot 11$
11	89	prime
12	144	$2^4 \cdot 3^2$
13	233	prime
14	377	$13 \cdot 29$
15	610	$2 \cdot 5 \cdot 61$
16	987	$3 \cdot 7 \cdot 47$
17	1597	prime
18	2584	$2^3 \cdot 17 \cdot 19$
19	4181	$37 \cdot 113$
20	6765	$3 \cdot 5 \cdot 11 \cdot 41$
21	10946	$2 \cdot 13 \cdot 421$
22	17711	$89 \cdot 199$
23	28657	prime
24	46368	$2^5 \cdot 3^2 \cdot 7 \cdot 23$
25	75025	$5^2 \cdot 3001$
26	121393	$233 \cdot 521$
27	196418	$2 \cdot 17 \cdot 53 \cdot 109$
28	317811	$3 \cdot 13 \cdot 29 \cdot 281$
29	514229	prime
30	832040	$2^3 \cdot 5 \cdot 11 \cdot 31 \cdot 61$
31	1346269	$557 \cdot 2417$
32	2178309	$3 \cdot 7 \cdot 47 \cdot 2207$
33	3524578	$2 \cdot 89 \cdot 19801$
34	5702887	$1597 \cdot 3571$
35	9227465	$5 \cdot 13 \cdot 141961$
36	14930352	$2^4 \cdot 3^3 \cdot 17 \cdot 19 \cdot 107$
37	24157817	$73 \cdot 149 \cdot 2221$
38	39088169	$37 \cdot 113 \cdot 9349$
39	63245986	$2 \cdot 233 \cdot 135721$
40	102334155	$3 \cdot 5 \cdot 7 \cdot 11 \cdot 41 \cdot 2161$

Figure 1-8

$F_6 = 8$
$F_8 = 21$
$F_9 = 34$
$F_{10} = 55$
$F_{12} = 144$
$F_{14} = 377$
$F_{15} = 610$
$F_{16} = 987$
$F_{18} = 2584$
$F_{20} = 6765$
$F_{21} = 10946$
$F_{22} = 17711$
$F_{24} = 46368$
$F_{25} = 75025$
$F_{26} = 121393$
$F_{27} = 196418$
$F_{28} = 317811$
$F_{30} = 832040$

Figure 1-9 Composite Fibonacci positions

$F_2 = 1$, not a prime.
$F_3 = 2$
$F_5 = 5$
$F_7 = 13$
$F_{11} = 89$
$F_{13} = 233$
$F_{17} = 1597$
$F_{19} = 4181 = 37 \cdot 113$
$F_{23} = 28657$
$F_{29} = 514229$

Figure 1-10 Prime number positioned Fibonacci numbers

Property 5

With all these lovely relationships embracing the Fibonacci numbers, there must be a simple way to get the sum of a specified number of these Fibonacci numbers. A simple formula would be more helpful than actually adding all the Fibonacci numbers to a certain point. To derive such a formula for the sum of the first n Fibonacci numbers, we will use a nice little technique that will help us generate a formula.

Remember the basic rule (or definition) of a Fibonacci number.

We can write this definition formally as $F_{n+2} = F_{n+1} + F_n$, where $n \geq 1$, or as $F_n = F_{n+2} - F_{n+1}$. By substituting increasing values for n, we get:

$$\begin{aligned} F_1 &= F_3 - F_2 \\ F_2 &= F_4 - F_3 \\ F_3 &= F_5 - F_4 \\ F_4 &= F_6 - F_5 \\ &\vdots \\ F_{n-1} &= F_{n+1} - F_n \\ F_n &= F_{n+2} - F_{n+1}. \end{aligned}$$

By adding these equations, you will notice that many terms on the right side of the equations will disappear because you will be adding and subtracting the same number and their sum is zero. What will remain on the right side will be $F_{n+2} - F_2 = F_{n+2} - 1$.

On the left side we have the sum of the first n Fibonacci numbers: $F_1 + F_2 + F_3 + F_4 + \cdots + F_n$, which is what we are looking for. Therefore, we get the following: $F_1 + F_2 + F_3 + F_4 + \cdots + F_n = F_{n+2} - 1$, which says that the sum of the first n Fibonacci numbers is equal to the Fibonacci number two further along the sequence minus 1. There is a shortcut notation that we can use to signify the sum: $F_1 + F_2 + F_3 + F_4 + \cdots + F_n$, and, that is, $\sum_{i=1}^{n} F_i$. This reads: "The sum of all the F_i terms where i takes on the values from 1 to n." So, we can then write this result as:

$$\sum_{i=1}^{n} F_i = F_1 + F_2 + F_3 + F_4 + \cdots + F_n = F_{n+2} - 1,$$

or simply:

$$\sum_{i=1}^{n} F_i = F_{n+2} - 1.$$

Property 6

Suppose we now consider the sum of the consecutive even-positioned Fibonacci numbers, beginning with the first such Fibonacci number, F_2. Again, let's see if we can discover a pattern when adding these consecutive even-positioned Fibonacci numbers.

$$F_2 + F_4 = 1 + 3 = 4$$

$$F_2 + F_4 + F_6 = 1 + 3 + 8 = 12$$

$$F_2 + F_4 + F_6 + F_8 = 1 + 3 + 8 + 21 = 33$$

$$F_2 + F_4 + F_6 + F_8 + F_{10} = 1 + 3 + 8 + 21 + 55 = 88.$$

We may notice that each sum is 1 less than a Fibonacci number—more particularly, the Fibonacci number that follows the last even number in the sum. We can write this symbolically as

$$F_2 + F_4 + F_6 + F_8 + \cdots + F_{2n-2} + F_{2n} = F_{2n+1} - 1, \quad \text{where } n \geq 1 \quad \text{or}$$

$$\sum_{i=1}^{n} F_{2i} = F_{2n+1} - 1, \quad \text{where } n \geq 1.$$

Property 7

Now that we have established a short-cut to get the sum of the initial consecutive even-positioned Fibonacci numbers, it is only fitting that we now inspect the analog: the sum of the initial consecutive odd-positioned Fibonacci numbers. Once again, by considering a few examples of these sums, we will look for a pattern.

$$F_1 + F_3 = 1 + 2 = 3$$

$$F_1 + F_3 + F_5 = 1 + 2 + 5 = 8$$

$$F_1 + F_3 + F_5 + F_7 = 1 + 2 + 5 + 13 = 21$$

$$F_1 + F_3 + F_5 + F_7 + F_9 = 1 + 2 + 5 + 13 + 34 = 55.$$

These sums appear to be Fibonacci numbers. But how do these sums relate to the series that generated them? In each case, the sums are the next

Fibonacci number after the last term in the sum of odd-positioned Fibonacci numbers. This can be symbolically written as:

$$F_1 + F_3 + F_5 + F_7 + \cdots + F_{2n-1} = F_{2n} \quad \text{or}$$

$$\sum_{i=1}^{n} F_{2i-1} = F_{2n}.$$

If we add the sum of the initial consecutive even-positioned Fibonacci numbers and the sum of the initial consecutive odd-positioned Fibonacci numbers, we should get the sum of the initial consecutive Fibonacci numbers:

$$F_2 + F_4 + F_6 + F_8 + \cdots + F_{2n} = F_{2n+1} - 1, \quad \text{where } n \geq 1,$$

$$F_1 + F_3 + F_5 + F_7 + \cdots + F_{2n-1} = F_{2n}, \quad \text{where } n \geq 1.$$

The sum of these sequences is:

$$F_1 + F_2 + F_3 + F_4 + \cdots + F_{2n} = F_{2n+1} - 1 + F_{2n}, \quad \text{or}$$

$$F_1 + F_2 + F_3 + F_4 + \cdots + F_{2n} = F_{2n+2} - 1,$$

which is consistent with what we concluded in item 4.

Property 8

Having established relationships for various sums of Fibonacci numbers, we shall now consider the sum of the *squares* of the initial Fibonacci numbers. Here we will see another astonishing relationship that continues to make the Fibonacci numbers special. Since we are talking about "squares," it is fitting for us to look at them geometrically.

We find that, beginning with a 1×1 square, as shown in Figure 1-11, we can generate a series of squares, the sides of which are Fibonacci numbers. We can continue on this way indefinitely. Let us now express the area of each rectangle as the sum of its component squares:

$$1^2 + 1^2 + 2^2 + 3^2 + 5^2 + 8^2 + 13^2 = 13 \cdot 21.$$

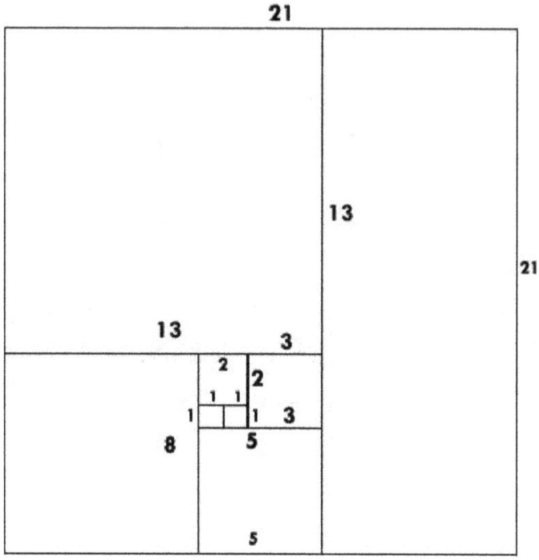

Figure 1-11 Squares of Fibonacci Numbers

If we express the sum of the squares making up the smaller rectangles, we get this pattern:

$$1^2 + 1^2 = 1 \cdot 2$$
$$1^2 + 1^2 + 2^2 = 2 \cdot 3$$
$$1^2 + 1^2 + 2^2 + 3^2 = 3 \cdot 5$$
$$1^2 + 1^2 + 2^2 + 3^2 + 5^2 = 5 \cdot 8$$
$$1^2 + 1^2 + 2^2 + 3^2 + 5^2 + 8^2 = 8 \cdot 13$$
$$1^2 + 1^2 + 2^2 + 3^2 + 5^2 + 8^2 + 13^2 = 13 \cdot 21.$$

From this pattern we can establish a rule: the sum of the squares of the Fibonacci numbers, to a certain point in the sequence, is equal to the product of the last number and the next number in the sequence. That is, if, for example, we choose to find the sum of the squares of the following portion of the sequence of Fibonacci numbers: 1, 1, 2, 3, 5, 8, 13, 21, 34, we would get:

$$1^2 + 1^2 + 2^2 + 3^2 + 5^2 + 8^2 + 13^2 + 21^2 + 34^2 = 1870.$$

However, this amazing rule tells us that this sum is merely the product of the last number and the one that would come after it (sometimes known as its immediate successor) in the Fibonacci sequence. That would mean that this sum can be found by multiplying 34 by 55. Indeed, this product gives us 1870 (= 34 · 55). We can write this with our summation notation as $\sum_{i=1}^{n}(F_i)^2 = F_n F_{n+1}$.

Imagine you want to find the sum of the squares of the first 30 Fibonacci numbers. This neat relationship makes the task very simple. Instead of finding the squares of each and then finding their sum—a rather laborious task—all we need to do is multiply the 30th by the 31st Fibonacci number. That is, the sum of the squares:

$$\sum_{i=1}^{30}(F_i)^2 = (F_1)^2 + (F_2)^2 + \cdots + (F_{29})^2 + (F_{30})^2$$
$$= 1^2 + 1^2 + 2^2 + 3^2 + 5^2 + 8^2 + 13^2 + \cdots + 514229^2 + 832040^2 = 1{,}120{,}149{,}658{,}760.$$

Or, simply using our newly established formula:

$$\sum_{i=1}^{30}(F_i)^2 = F_{30} \cdot F_{31} = 832040 \cdot 1346269 = 1{,}120{,}149{,}658{,}760.$$

Property 9

While we are talking about the squares of Fibonacci numbers, consider the following relationship. Let's take the square of a Fibonacci number, say, 34, and the square of the Fibonacci number that is two before it, 13, and then subtract the two squares. We get $34^2 - 13^2 = 1156 - 169 = 987$, which is also a Fibonacci number! So, we squared the 9th Fibonacci number ($F_9 = 34$), and subtracted the square of the 7th Fibonacci number ($F_7 = 13$) to get the 16th Fibonacci number ($F_{16} = 987$). This can be written symbolically as: $F_9^2 - F_7^2 = F_{16}$.

It is nice to see that by subtracting the squares of two alternating[11] Fibonacci numbers, we seem to get another Fibonacci number. Will this work for any pair of alternating Fibonacci numbers? We can begin to convince ourselves by inspecting some examples to see that this result was no fluke, but actually works for any appropriate pair of Fibonacci numbers. Let's look at a few now and see if we can determine a pattern:

$$F_6{}^2 - F_4{}^2 = 8^2 - 3^2 = 55 = F_{10}$$

$$F_7{}^2 - F_5{}^2 = 13^2 - 5^2 = 144 = F_{12}$$

$$F_{15}{}^2 - F_{13}{}^2 = 610^2 - 233^2 = 317811 = F_{28}.$$

By inspecting the three subscripts for each of these examples, we notice that the sum of the first two equals the third. This would lead us to make the following generalization: $F_n^2 - F_{n-2}^2 = F_{2n-2}$. You might want to try to apply this rule to a few more such pairs to further convince yourself.

Property 10

The natural next step would be to look at the sum of the squares of two Fibonacci numbers. Consider two consecutive Fibonacci numbers, say, $F_7(=13)$ and $F_8(=21)$. Their squares, 169 and 441, have a sum of 610, which is also a Fibonacci number, F_{15}. Perhaps we are onto something here. Suppose we try another consecutive pair of Fibonacci numbers and see what the sum of the squares of these numbers is. To establish a pattern, we will look at the two Fibonacci numbers F_{10} (=55) and F_{11} (=89). Their squares, 3025 and 7921, have a sum of 10946, which is again a Fibonacci number. It appears that when we try this for several other pairs of consecutive Fibonacci numbers, we arrive each time at another Fibonacci number. But what is the pattern? To predict which Fibonacci number will result from the sum of the squares of a consecutive pair of Fibonacci numbers, we will want to inspect their position in the sequence. In our first example, above, we used the Fibonacci numbers F_7, and F_8, and the sum of their squares was F_{15}. In the next example we used the Fibonacci numbers F_{10}, and F_{11},

---

[11] Alternating members of a sequence will be those that are one member apart from one another. For example, the 4th and 6th members or the 15th and 17th members are alternating members.

and we found that the sum of their squares was F_{21}. It appears that the sum of the subscripts of the two Fibonacci numbers whose squares we are adding will give us the subscript of the Fibonacci number representing their sum. That is, the sum of the squares of the Fibonacci numbers in positions n and $n + 1$ (consecutive positions) is the Fibonacci number in place $n + (n + 1) = 2n + 1$, or $F_n^2 + {F_{n+1}}^2 = F_{2n+1}$.

Property 11

Here is an engaging relationship that takes another look at the unexpected patterns we find among the numbers in the Fibonacci sequence. Take any four consecutive numbers in the sequence, say, 3, 5, 8, 13, and then find the difference of the squares of the middle two numbers: $8^2 - 5^2 = 64 - 25 = 39$. Then find the product of the outer two numbers: $3 \cdot 13 = 39$. Amazingly the same result is attained! Was this by strange coincidence or was it a pattern that will hold for all such strings of four consecutive Fibonacci numbers? We can try this again with another group of four consecutive Fibonacci numbers, say, 8, 13, 21, 34. Again, we find the sum of the squares of the two middle numbers: $21^2 - 13^2 = 441 - 169 = 272$. If the pattern will continue to work, then the product of the outer two numbers must be 272. Well, $8 \cdot 34 = 272$, and so the pattern still holds. Symbolically, we can write this as $(F_{n+1})^2 - (F_n)^2 = F_{n-1} \cdot F_{n+2}$. To really convince yourself that this is always true try many groups of four consecutive Fibonacci numbers and find that the pattern holds for each of your tries.

Property 12

Another curious relationship is obtained by inspecting the products of two alternating members of the Fibonacci sequence. Consider some of these products:

- $F_3 \cdot F_5 = 2 \cdot 5 = 10$, which is 1 more than the square of the Fibonacci number 3: $(F_4)^2 + 1 = 3^2 + 1$;
- $F_4 \cdot F_6 = 3\cdot = 24$, which is 1 less than the square of the Fibonacci number 5: $(F_5)^2 - 1 = 5^2 - 1$;
- $F_5 \cdot F_7 = 5 \cdot 13 = 65$, which is 1 more than the square of the Fibonacci number 8: $(F_6)^2 + 1 = 8^2 + 1$;

- $F_6 \cdot F_8 = 8 \cdot 21 = 168$, which is 1 less than the square of the Fibonacci number 13: $(F_7)^2 - 1 = 13^2 - 1$.

By now you should begin to see a pattern emerging: The product of two alternating Fibonacci numbers is equal to the square of the Fibonacci number between them ± 1. We still need to determine when it is $+1$ or -1. When the number to be squared is in an even-numbered position, we add 1, and when it is in an odd-numbered position, we subtract 1. This can be generalized by using $(-1)^n$, since that will do exactly what we want: -1 to an even power is $+1$, and -1 to an odd power is -1. Symbolically this may be written as: $F_{n-1}F_{n+1} = F_n^2 + (-1)^n$, where $n \geq 1$.

This relationship can be expanded. Suppose instead of taking the product of the two Fibonacci numbers on either side of a particular Fibonacci number as we did above, we would take the two Fibonacci numbers that are *one* removed in either direction. Let's see how the product compares to the square of the Fibonacci number in the middle if we take a specific example from the Fibonacci numbers, say, $F_6 = 8$. The product of the two Fibonacci numbers one removed on either side of 8 is $3 \cdot 21 = 63$, and the square of 8 is 64. They differ by 1. Suppose we now use $F_5 = 5$, then the product of the two number ones removed on either side is $2 \cdot 13 = 26$, which differs from 5^2 by 1. We can write this symbolically as $F_{n-2}F_{n+2} = F_n^2 \pm 1$, where $n \geq 1$.

We can now try this by comparing the product of Fibonacci numbers two, three, four, etc. removed on either side of a designated Fibonacci number and you will find the following to hold true.

By now you may begin to recognize the pattern in the chart in Figure 1-12. The difference between the square of the selected Fibonacci number and the various products of Fibonacci numbers, which are equidistant from the selected Fibonacci number, is the square of another Fibonacci number. Symbolically we can write this as $F_{n-k}F_{n+k} - F_n^2 = \pm F_k^2$, where $n \geq 1$, and $k \geq 1$.

Property 13

There are endless properties that one can observe with the Fibonacci numbers. Many of them can be found by simple inspection. Let's take

Number of Fibonacci numbers removed on either side of selected Fibonacci number	Symbolic representation for the case of F_n		Example for $F_7 = 13$		Example for $F_8 = 21$		Difference between: $F_{n-k}F_{n+k}$ and F_n^2
1	$F_{n-1}F_{n+1}$	F_n^2	$8\cdot 21 = 168$	$13^2 = 169$	$13\cdot 34 = 442$	$21^2 = 441$	± 1
2	$F_{n-2}F_{n+2}$	F_n^2	$5\cdot 34 = 170$	$13^2 = 169$	$8\cdot 55 = 440$	$21^2 = 441$	± 1
3	$F_{n-3}F_{n+3}$	F_n^2	$3\cdot 55 = 165$	$13^2 = 169$	$5\cdot 89 = 445$	$21^2 = 441$	± 4
4	$F_{n-4}F_{n+4}$	F_n^2	$2\cdot 89 = 178$	$13^2 = 169$	$3\cdot 144 = 432$	$21^2 = 441$	± 9
5	$F_{n-5}F_{n+5}$	F_n^2	$1\cdot 144 = 144$	$13^2 = 169$	$2\cdot 233 = 466$	$21^2 = 441$	± 25
6	$F_{n-6}F_{n+6}$	F_n^2	$1\cdot 233 = 233$	$13^2 = 169$	$1\cdot 377 = 377$	$21^2 = 441$	± 64
k	$F_{n-k}F_{n+k}$	F_n^2					$\pm F_k^2$

Figure 1-12 Unusual Fibonacci Relationships

another look at the sequence in Figure 1-13. Beginning with the first Fibonacci number, notice that every third number is even. That is, $F_3, F_6, F_9, F_{12}, F_{15}$, and F_{18} are all even numbers. We could restate this by saying that these numbers are all divisible by 2 or by F_3.

Let's further examine this list of Fibonacci numbers. Notice that every fourth number is divisible by 3. Here, $F_4, F_8, F_{12}, F_{16}, F_{20}, F_{24}$ are each divisible[12] by 3. We can again restate this by saying that $F_4, F_8, F_{12}, F_{16}, F_{20}, F_{24}$ are each divisible by F_4, or by 3.

Using patterns to see where we go from here, we would see that the first divisibility was by 2, then by 3, then, in good Fibonacci style, we ought to try to check for divisibility by 5—the next number in the Fibonacci sequence. Searching for Fibonacci numbers that are divisible by 5 is easy. They are the numbers 5, 55, 610, 6765, 75025, 832040, . . . , which correspond (symbolically) to $F_5, F_{10}, F_{15}, F_{20}, F_{25}, F_{30}$. Thus, we have that $F_5, F_{10}, F_{15}, F_{20}, F_{25}, F_{30}$ are each divisible by 5 or by F_5.

[12]You can check this quite easily by using the popular rule for divisibility by 3. If and only if the sum of the digits of a number is divisible by 3, then the number itself is divisible by 3.

F_1	1
F_2	1
F_3	2
F_4	3
F_5	5
F_6	8
F_7	13
F_8	21
F_9	34
F_{10}	55
F_{11}	89
F_{12}	144
F_{13}	233
F_{14}	377
F_{15}	610
F_{16}	987
F_{17}	1597
F_{18}	2584
F_{19}	4181
F_{20}	6765
F_{21}	10946
F_{22}	17711
F_{23}	28657
F_{24}	46368
F_{25}	75025
F_{26}	121393
F_{27}	196418
F_{28}	317811
F_{29}	514229
F_{30}	832040

Figure 1-13 Fibonacci Numbers

Checking for divisibility by 8 (the next Fibonacci number), we find that $F_6, F_{12}, F_{18}, F_{24}, F_{30}$ are each divisible by 8, or by F_6. Yes, every seventh Fibonacci number is divisible by 13, or F_7. You might now try to generalize this finding. You can either say that a Fibonacci number F_{nm} is divisible by a Fibonacci number F_m, where n is a positive integer, or you may state it as the following: If p is divisible by q, then F_p is divisible by F_q.

Property 14

Fibonacci relationships, as noted earlier, can also be seen geometrically. We consider Fibonacci squares as arranged as in Figure 1-14 and Figure 1-15.

Figure 1-14 shows an odd number ($n = 7$) of rectangles into which the Fibonacci square is divided. The rectangles have dimensions of F_i and F_{i+1}, where i has values from 1 to $n + 1$. The square then has sides of length F_{n+1}. The area of the square is equal to the sum of the rectangles, which can constitute a geometric proof. This can be written symbolically as $\sum_{i=2}^{n+1} F_i F_{i-1} = F_{n+1}^2$, when n is odd.

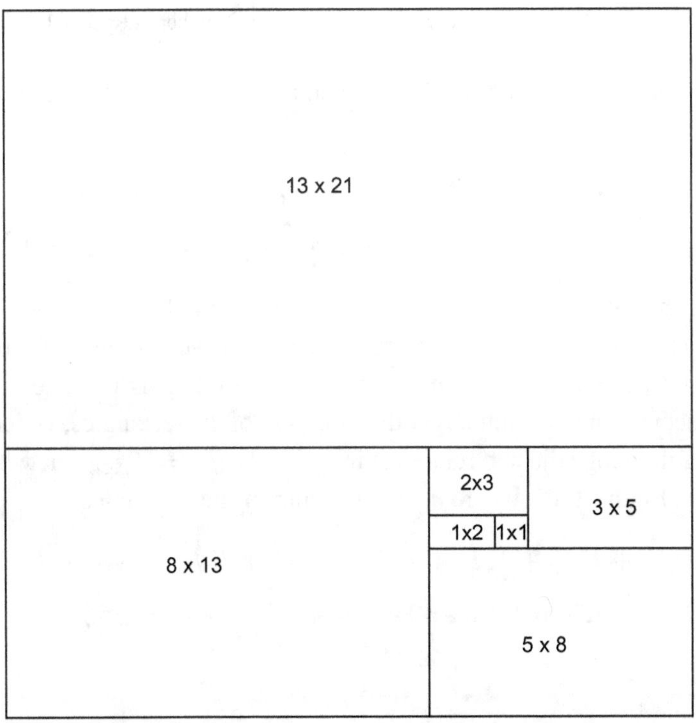

Figure 1-14 Square with an *Odd* Number of Rectangles

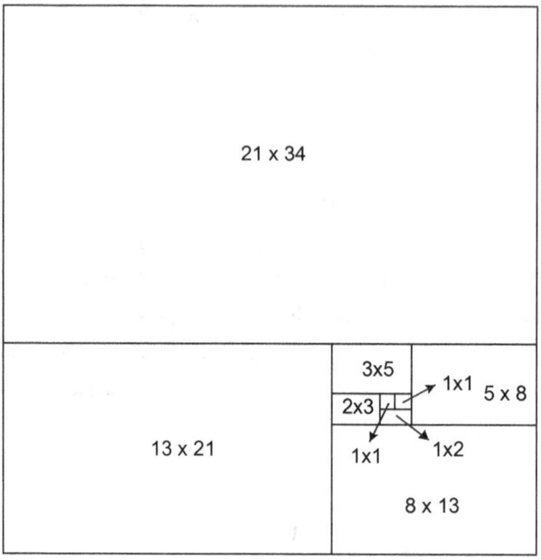

Figure 1-15 Fibonacci Square with an *Even* Number of Rectangles

As you can see in Figure 1-14, when $n = 7$ the sum of the areas of the rectangles is

$$F_2F_1 + F_3F_2 + F_4F_3 + F_5F_4 + F_6F_5 + F_7F_6 + F_8F_7$$
$$= 1 + 2 + 6 + 15 + 40 + 104 + 273 = 441 = 21^2 = F_8^2.$$

On the other hand, when n is even (in Figure 1-15, $n = 8$), that is, when a Fibonacci square is constructed using an even number of Fibonacci rectangles, a unit square remains that is not used. This indicates that the Fibonacci square is 1 unit larger than the sum of the rectangles, which gives us the following adjusted relationship $\sum_{i=2}^{n+1} F_i F_{i-1} = F_{n+1}^2 - 1$, when n is even. In Figure 1-15, where $n = 8$, the sum of the rectangles is

$$1 + F_2F_1 + F_3F_2 + F_4F_3 + F_5F_4 + F_6F_5 + F_7F_6 + F_8F_7 + F_9F_8$$
$$= 1 + 1 + 2 + 6 + 15 + 40 + 104 + 273 + 714 = 1156 = 34^2 = F_9^2.$$

Property 15

Earlier in this chapter, we mentioned the Lucas numbers. They are: 1, 3, 4, 7, 11, 18, 29, 47, ... We can find the sum of the Lucas numbers much the same way we found the sum of the Fibonacci numbers. Again, there must be a simple formula to get the sum of a specified number of these Lucas numbers. To derive such a formula for the sum of the first n Lucas numbers, we will use a nice little technique that will help us generate the formula.

The basic rule (or definition) of a Lucas number is $L_{n+2} = L_{n+1} + L_n$, where $n \geq 1$. Or $L_n = L_{n+2} - L_{n+1}$

By substituting increasing values for n, we get:

$$\begin{aligned} L_1 &= L_3 - L_2 \\ L_2 &= L_4 - L_3 \\ L_3 &= L_5 - L_4 \\ L_4 &= L_6 - L_5 \\ &\vdots \\ L_{n-1} &= L_{n+1} - L_n \\ L_n &= L_{n+2} - L_{n+1} \end{aligned}$$

When adding these equations, you will notice that many terms on the right side will disappear because their sum is zero. This occurs because you will be adding and subtracting the same number. What will remain on the right side will be $L_{n+2} - L_2 = L_{n+2} - 3$.

On the left side we have the sum of the first n Lucas numbers $L_1 + L_2 + L_3 + L_4 + \cdots + L_n$, which is what we are looking for. Therefore, we get the following: $L_1 + L_2 + L_3 + L_4 + \cdots + L_n = L_{n+2} - 3$, which says that the sum of the first n Lucas numbers is equal to the Lucas number two further along the sequence minus 3.

There is a shortcut notation that we can use to signify the sum $L_1 + L_2 + L_3 + L_4 + \cdots + L_n$, and that is $\sum_{i=1}^{n} L_i$. We can write this result as $\sum_{i=1}^{n} L_i = L_1 + L_2 + L_3 + L_4 + \cdots + L_n = L_{n+2} - 3$, or simply $\sum_{i=1}^{n} L_i = L_{n+2} - 3$.

Property 16

Just as we found the sum of the squares of the Fibonacci numbers, so, too, can we find the sum of the squares of the Lucas numbers.

Here we will uncover another astonishing relationship that makes the Lucas numbers special.

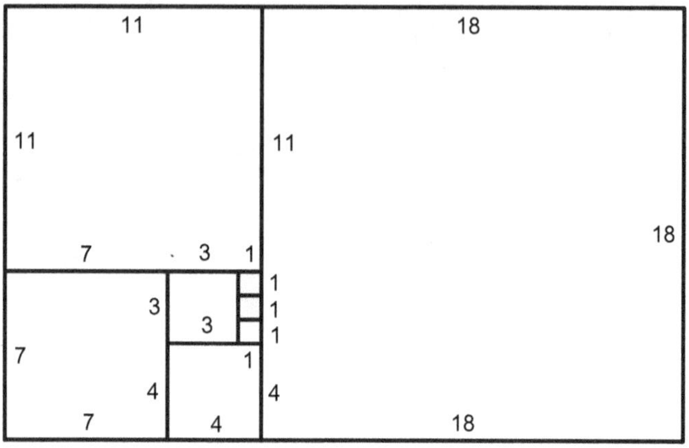

Figure 1-16 Sum of Squares of Lucas Numbers

In Figure 1-16, beginning with the three small 1 × 1 squares, we can generate a series of squares the sides of which are Lucas numbers. We can continue on this way indefinitely. Let us now express the area of the rectangle as the sum of its component squares minus the two small squares (shaded).

$$L_1^2 + L_2^2 + L_3^2 + L_4^2 + L_5^2 + L_6^2 = 1^2 + 3^2 + 4^2 + 7^2 + 11^2 + 18^2$$
$$= 522 - 2 = 520 = 18 \cdot 29 = L_6 \cdot L_7.$$

If we express the sum of the squares progressively, which comprise the smaller rectangles, we get this pattern:

$$1^2 + 3^2 = 3 \cdot 4 - 2$$
$$1^2 + 3^2 + 4^2 = 4 \cdot 7 - 2$$
$$1^2 + 3^2 + 4^2 + 7^2 = 7 \cdot 11 - 2$$
$$1^2 + 3^2 + 4^2 + 7^2 + 11^2 = 11 \cdot 18 - 2$$
$$1^2 + 3^2 + 4^2 + 7^2 + 11^2 + 18^2 = 18 \cdot 29 - 2$$
$$1^2 + 3^2 + 4^2 + 7^2 + 11^2 + 18^2 + 29^2 = 29 \cdot 47 - 2.$$

From this pattern we can establish a rule: the sum of the squares of the Lucas numbers to a certain point in the sequence is equal to the product of the last number and the next number in the sequence minus 2. If, for example, we choose to find the sum of the squares of the following portion of the sequence of Lucas numbers 1, 3, 4, 7, 11, 18, 29, 47, 76, we would get

$$L_1^2 + L_2^2 + L_3^2 + L_4^2 + L_5^2 + L_6^2 + L_7^2 + L_8^2 + L_9^2$$
$$= 1^2 + 3^2 + 4^2 + 7^2 + 11^2 + 18^2 + 29^2 + 47^2 + 76^2 = 9346.$$

However, we can apply this amazing rule that tells us that this sum is merely the product of the last number and the one that would come after it (sometimes known as its immediate successor) in the Lucas sequence. That would mean that this sum can be found by multiplying 76 by 123 and subtracting 2. Indeed, this product gives us $L_9 \cdot L_{10} = 76 \cdot 123 = 9348 = 9346 + 2$. We can concisely summarize this as $\sum_{i=1}^{n} (L_i)^2 = L_n L_{n+1} - 2$. This will help you find the sum of the squares of the Lucas numbers without having to first find all the terms below it.

Imagine you would want to find the sum of the squares of the first 30 Lucas numbers. This neat relationship makes the task very simple. Instead of finding the squares of each and then finding their sum—a time consuming task—all you need to do is multiply the 30th by the 31st Lucas number and subtract 2. These proofs, although geometric, are reasonable enough to convince you of the veracity of these statements. Before you explore more unexpected sightings of these lovely Lucas numbers, we provide you with a summary of the properties we have noted thus far.

Summary of the Properties

Here is a summary of the Fibonacci (and Lucas) relationships we have seen (Let be n any natural number; $n \geq 1$). A definition of the Fibonacci numbers F_n and of the Lucas numbers L_n is as follows:

$$F_1 = 1; \quad F_2 = 1; \quad F_{n+2} = F_n + F_{n+1}$$
$$L_1 = 1; \quad L_2 = 3; \quad L_{n+2} = L_n + L_{n+1}$$

Property 1. *The sum of any ten consecutive Fibonacci numbers is divisible by 11*: $11 \mid (F_n + F_{n+1} + F_{n+2} + \cdots + F_{n+8} + F_{n+9})$.

Property 2. *Consecutive Fibonacci numbers are relatively prime*. (That is, their greatest common divisor is 1.)

Property 3. *Fibonacci numbers in a composite number position (with the exception of the 4*th *Fibonacci number) are also composite numbers.* (Another way of saying this is that if n is not prime then F_n is not prime (with $n \neq 4$, since $F_4 = 3$, which is a prime number).)

Property 4. *The sum of the first n Fibonacci numbers is equal to the Fibonacci number two further along the sequence minus 1*: $\sum_{i=1}^{n} F_i = F_1 + F_2 + F_3 + F_4 + \cdots + F_n = F_{n+2} - 1$.

Property 5. *The sum of the consecutive even-positioned Fibonacci numbers is 1 less than the Fibonacci number that follows the last even number in the sum*: $\sum_{i=1}^{n} F_{2i} = F_2 + F_4 + F_6 + \cdots + F_{2n-2} + F_{2n} = F_{2n+1} - 1$.

Property 6. *The sum of the consecutive odd-positioned Fibonacci numbers is equal to the Fibonacci number that follows the last odd number in the sum*: $\sum_{i=1}^{n} F_{2i-1} = F_1 + F_3 + F_5 + \cdots + F_{2n-3} + F_{2n-1} = F_{2n}$.

Property 7. *The sum of the squares of the Fibonacci numbers is equal to the product of the Fibonacci numbers of the last number and the next number in the sequence*: $\sum_{i=1}^{n} (F_i)^2 = F_n F_{n+1}$.

Property 8. *The difference of the squares of two alternate*[13] *Fibonacci numbers is equal to the Fibonacci number in the sequence whose position number is the sum of their position numbers*:

$$F_n^2 - F_{n-2}^2 = F_{2n-2}.$$

Property 9. *The sum of the squares of two consecutive Fibonacci numbers is equal to the Fibonacci number in the sequence whose position number is the sum of their position numbers*:

$$F_n{}^2 + F_{n+1}{}^2 = F_{2n+1}.$$

Property 10. *For any group of four consecutive Fibonacci numbers the difference of the squares of the middle two numbers is equal to the product*

---

[13] i.e., a Fibonacci number with number n and the Fibonacci number two before it.

of the outer two numbers. (Symbolically, we can write this as $F_{n+1}^2 - F_n^2 = F_{n-1} \cdot F_{n+2}$.)

Property 11. *The product of two alternating Fibonacci numbers is 1 more or less than the square of the Fibonacci number between them* ± 1: $F_{n-1}F_{n+1} = F_n^2 + (-1)^n$. (If n is even, the product is 1 more; if n is odd, the product is 1 less.)

Property 12. *The difference between the square of the selected Fibonacci number and the various products of Fibonacci numbers equidistant from the selected Fibonacci number is the square of another Fibonacci number:* $F_{n-k}F_{n+k} - {F_n}^2 = \pm F_k^2$, where $n \geq 1$, and $k \geq 1$.

Property 13. *A Fibonacci number F_{mn} is divisible by a Fibonacci number F_m*. (We can write this as $F_m | F_{mn}$, and it reads "F_m divides F_{mn}." In other words: If p is divisible by q, then F_p is divisible by F_q. Using our symbols: $q|p \Rightarrow F_q|F_p (m, n, p, q$ are positive integers).)

Here is how this looks for specific cases:

$$
\begin{aligned}
&[F_1|F_n, && \text{i.e.,} && 1|F_1, 1|F_2, 1|F_3, 1|F_4, 1|F_5, 1|F_6, \ldots, 1|F_n, \ldots \\
&F_2|F_{2n}, && \text{i.e.,} && 1|F_2, 1|F_4, 1|F_6, 1|F_8, 1|F_{10}, 1|F_{12}, \ldots, 1|F_{2n}, \ldots] \\
&F_3|F_{3n}, && \text{i.e.,} && 2|F_3, 2|F_6, 2|F_9, 2|F_{12}, 2|F_{15}, 2|F_{18}, \ldots, 2|F_{3n}, \ldots \\
&F_4|F_{4n}, && \text{i.e.,} && 3|F_4, 3|F_8, 3|F_{12}, 3|F_{16}, 3|F_{20}, 3|F_{24}, \ldots, 3|F_{4n}, \ldots \\
&F_5|F_{5n}, && \text{i.e.,} && 5|F_5, 5|F_{10}, 5|F_{15}, 5|F_{20}, 5|F_{25}, 5|F_{30}, \ldots, 5|F_{5n}, \ldots \\
&F_6|F_{6n}, && \text{i.e.,} && 8|F_6, 8|F_{12}, 8|F_{18}, 8|F_{24}, 8|F_{30}, 8|F_{36}, \ldots, 8|F_{6n}, \ldots \\
&F_7|F_{7n}, && \text{i.e.,} && 13|F_7, 13|F_{14}, 13|F_{21}, 13|F_{28}, 13|F_{35}, 13|F_{42}, \ldots, \\
& && && 13|F_{7n}, \ldots
\end{aligned}
$$

Property 14. *The product of consecutive Fibonacci numbers is either the square of a Fibonacci number or 1 greater that the square of a Fibonacci number.*

$$\sum_{i=2}^{n+1} F_i F_{i-1} = F_{n+1}^2, \quad \text{when } n \text{ is odd.}$$

$$\sum_{i=2}^{n+1} F_i F_{i-1} = F_{n+1}^2 - 1, \quad \text{when } n \text{ is even.}$$

Property 15. *The sum of the first n Lucas numbers is equal to the Lucas number two further along the sequence minus 3*: $\sum_{i=1}^{n} L_i = L_1 + L_2 + L_3 + L_4 + \cdots + L_n = L_{n+2} - 3$.

Property 16. *The sum of the squares of the Lucas numbers is equal to the product of the Lucas numbers of the last number and the next number in the sequence minus 2:* $\sum_{i=1}^{n} (L_i)^2 = L_n L_{n+1} - 2$.

Although our focus is largely on the Fibonacci numbers, we should not think of Fibonacci as being known only for this now-famous sequence of numbers that bears his name. Fibonacci, as mentioned, was one of the greatest mathematicians in Western culture. He not only provided us with the tools to perform efficient mathematics (e.g., he introduced the Hindu-Arabic numerals theretofore unknown to Europe) but he also introduced to the European world a thought process that opened the way for many future mathematical endeavors.

Notice how the Fibonacci numbers, innocently embedded in a problem about the regeneration of rabbits, seem to have properties far beyond what might be expected. The surprising relationships of this sequence, within the realm of the numbers, are truly mind-boggling! It is this phenomenon coupled with the almost endless applications beyond the sequence that has intrigued mathematicians for generations. It is our intention to fascinate you with applications of these numbers as far afield as one can imagine.

Chapter 2

Fibonacci Numbers in Everyday Occurrences

Fibonacci numbers can be found in the most unexpected places. The Fibonacci numbers are closely related to the Golden Ratio, and appear everywhere in everyday life, even in seemingly nonmathematical occurrences. That is what we will consider in this chapter.

The Vending Machines

Suppose you are faced with the task of setting up a vending machine to dispense a variety of candies. The cost of each of the candies should be a multiple of 25 cents. Each type of candy will require a different arrangement (order) of coins deposited so that the amount of money deposited into the machine and the order in which the coins have been deposited will determine which of the candies has been selected. We need to calculate how many different types of candy the machine can accommodate. Put another way, we need to determine the number of different ways each multiple of 25 cents can be arrived at using the machine's coin slot, which accepts only quarters and half dollars. For example, for a 75-cent candy selection, there are three possible ways to pay: three quarters (QQQ), a quarter and a half dollar (QH), or a half dollar and a quarter (HQ). Each of these payments will result in a different type of candy. When we make a chart, shown in Figure 2-1, of the various amounts of money that the machine can accept, a curious set of numbers appears. Can you guess what that might be?

Cost of the candies	Number of multiples of 25 cents	List of ways to pay	Number of ways to pay
$.25	1	Q	1
$.50	2	QQ, H	2
$.75	3	QQQ, HQ, QH	3
$1.00	4	QQQQ, HH, QQH, HQQ, QHQ	5
$1.25	5	QQQQQ, HHQ, QHH, HQH, QQQH, HQQQ, QHQQ, QQHQ,	8
$1.50	6	QQQQQQ, QQQQH, QQQHQ, QQHQQ, QHQQQ, HQQQQ, QQHH, HHQQ, HQHQ, QHQH, HQQH, QHHQ, HHH,	13
$1.75	7	QQQQQQQ, QQQQQH, QQQQHQ, QQQHQQ, QQHQQQ, QHQQQQ, HQQQQQ, QQQHH, QQHHQ, QHHQQ, HHQQQ, QHQHQ, HQHQQ, QQHQH, HQQQH, HQQHQ, QHQQH, HHHQ, HHQH, HQHH, QHHH,	21
$2.00	8	QQQQQQQQ, QQQQQQH, QQQQQHQ, QQQQHQQ, QQQHQQQ, QQHQQQQ, QHQQQQQ, HQQQQQQ, QQQQHH, QQQHHQ, QQHHQQ, QHHQQQ, HHQQQQ, HQHQQQ, QHQHQQ, QQHQHQ, QQQQHH, HQQHQQ, QHQQHQ, QQHQQH, HQQQHQ, QHQQQH, HQQQQH, QQHHH, HQQHH, HHQQH, HHHQQ, QHQHH, HQHQH, HHQHQ, QHHQH, HQHHQ, QHHHQ, HHHH,	34

Figure 2-1 Coins in a vending machine

Once again, the Fibonacci numbers appear in the right-side column, indicating the number of ways by which the various amounts of money can be deposited. This would help the manufacturer of the vending machine predict the number of ways that one could deposit $3.00. There are 12 multiples of 25 cents in $3.00, and the answer would be the 13th number in the Fibonacci sequence, namely, 233. A knowledge of the Fibonacci numbers can be quite helpful.

Climbing a Staircase

Who would ever imagine that the Fibonacci numbers would be involved in the ways that we can climb a staircase? Let's investigate where the Fibonacci numbers might be hiding. Suppose we are about to climb a staircase with n stairs, and we have the option of either taking one step at a time or two steps at a time. We will now show how the number of different ways of climbing the stairs (C_n) will be a Fibonacci number. Suppose $n = 1$, which means there is only one stair to climb. Then the answer is easy: There is only **1** way to climb the stairs. If $n = 2$, and there are two stairs, then there are **2** different ways to climb the steps: two single steps or one double step. If $n = 3$, then with three steps there are **3** different ways to climb the steps: (1 step + 1 step + 1 step), (1 step + 2 steps), or (2 steps + 1) step. We can take that further to a staircase with four steps, where $n = 4$ and show that there are five different ways to climb the staircase, namely, (1 step + 1 step + 2 steps), (1 step + 2 steps + 1 step), (2 steps + 1 step + 1 step), (2 steps + 2 steps), (1 step + 1 step + 1 step + 1 step). By now you will have noticed the emergence of the Fibonacci numbers 1, 2, 3, and 5.

This can be generalized for the staircase of n steps: where $n > 2$ the number of ways of climbing them is C_n. After you take the first step up on a single stair, there are $n - 1$ stairs left to climb. So, the number of ways to climb the remainder of the staircase is C_{n-1}. If you begin your climb with a double step (i.e., 2 steps have been climbed), then there are C_{n-2} ways to climb the rest of the staircase. Thus, the number of ways to climb the staircase of n steps is the sum of these: $C_n = C_{n-1} + C_{n-2}$. From this you should recall that a Fibonacci number is the sum of its two predecessor numbers, which is the case shown here. The values of C_n will follow the Fibonacci pattern, with $C_n = F_{n+1}$. You can check this with smaller numbers of stairs and see that it works, or you can compare this to the previous vending machine problem and see that the two examples are equivalent, with the 25-cent units analogous to the stairs.

On the other hand, we can inspect a rather large example. The Empire State Building is 1,454 feet tall from street level to its highest point—a lightning rod. There are annual races up the building's 1,860 stairs. In how many ways can an Empire State Building stairs runner reach the top-most step? It is an unbelievably large number: $C_n = C_{1860} = F_{1861} =$ 377149471124318143225077447499310496327976870086234808713516 097645681561933736801512324129452985171904258339368239422753 956808208965187321202688520368618676247281289202395090152176 154315717419682601464319012327504645309682967175448664754029 173203923520902436572243276571313259547805808438502836830547 141311363286744699164434648027389766626163251643066565445211 33547290540333738912142760761 $\approx 3.771494711 \times 10^{388}$.

Painting a House Creatively

The levels of an n-level house are to be painted blue and yellow, with the restriction that no two adjacent levels are to be painted blue. However, adjacent levels can be painted yellow. We will let a_n represent the number of possible colorings for a house with n levels ($n \geq 1$).

In Figure 2-2, we show the possible ways a house of this various levels can be painted with the above restrictions. Now suppose we have a building with five levels. The fifth level can be painted either yellow or blue. If it gets a yellow color, then the remaining levels can be painted as a four-level house. If it is painted blue, the fourth level must have been painted yellow and the first three levels can be painted as a three-level house. Therefore, we find that $a_5 = a_4 + a_3 = 8 + 5 = 13 = F_6 + F_5 = F_7$. Once again, we see the Fibonacci numbers emerging when least expected.

Number of levels	Number of possibilities	Example
1	2	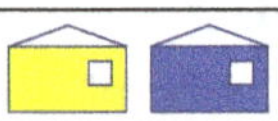
2	3	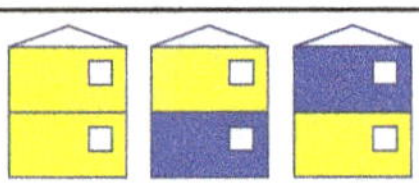
3	5	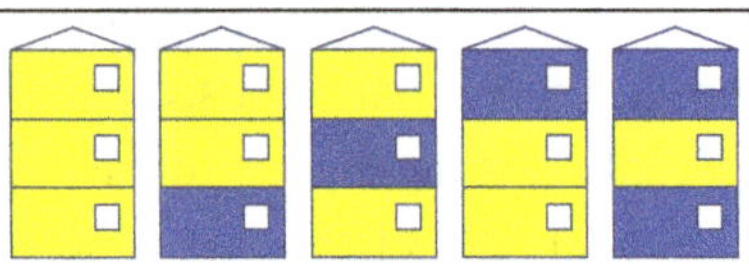
4	8	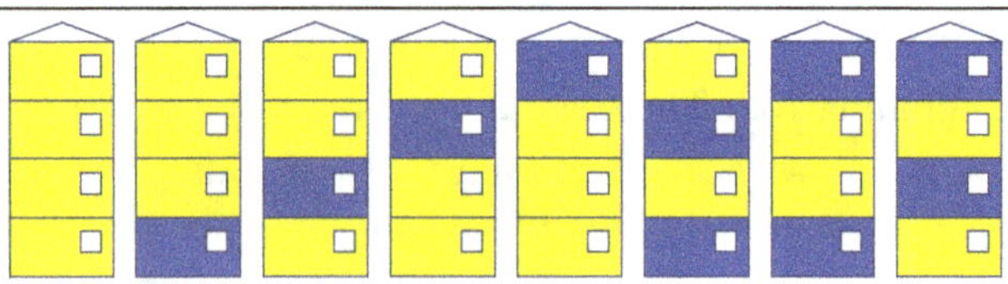

Figure 2-2 House Painting Options

Covering a Checkerboard

Most people are familiar with two games: dominoes and chess. Covering chess-type boards with domino-type tiles can reveal a surprising appearance of the Fibonacci numbers. We will use domino tiles that are the size of two adjacent cells on the chessboard. Furthermore, the chessboards that we will be using have two rows of pairs of cells and increasingly more columns. Mathematically, we would say that the tiles are 2×1 in size and the chessboards are increasingly $2 \times (n - 1)$ in size. The smallest of these chessboards, shown in Figure 2-3, has only one column, and therefore can only be covered with the domino in one way.

When we increase the size of the chessboard to have two columns, as we show in Figure 2-4, there are two ways to cover the enlarged chessboard.

Enlarging the chessboard to three columns, as we do in Figure 2-5, we find that there are three ways to cover the chessboard with the dominoes.

The next enlargement of the chessboard would have four columns and can be covered in five different ways, as we show in Figure 2-6.

Moving along to a chessboard with 5 columns, we find that there are 8 ways to cover the board with the dominoes, as shown in Figure 2-7.

Figure 2-3

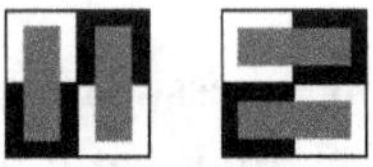

Figure 2-4

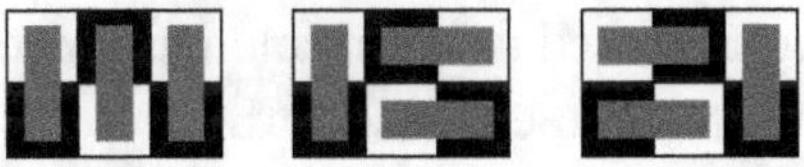

Figure 2-5

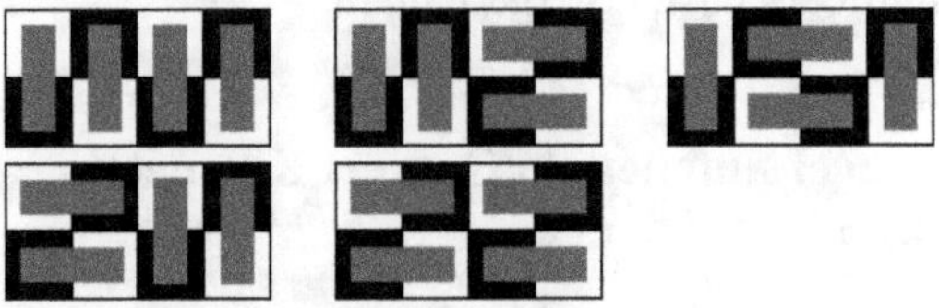

Figure 2-6

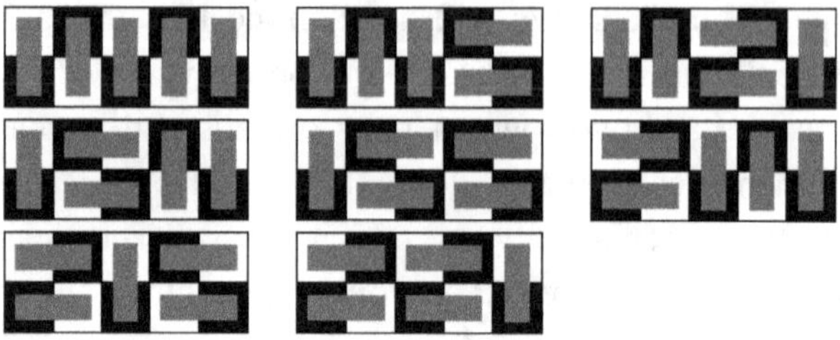

Figure 2-7

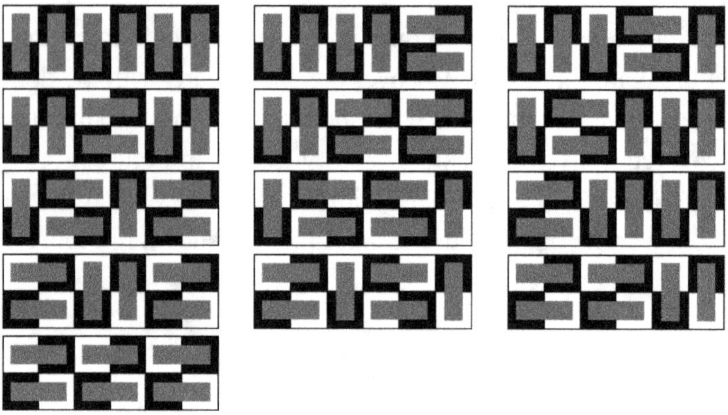

Figure 2-8

Once again, we increase the chessboard by one column to get a chessboard with six columns and we find that it can be covered in 13 different ways, as shown in Figure 2-8.

To further support the appearance of the Fibonacci numbers, consider the chessboard being increased to seven columns. We can see in Figure 2-9 that there are 21 ways to tile the chessboard.

Looking back now at Figures 2-3–2-9, we find that the number of ways to cover these truncated chessboards are 1, 2, 3, 5, 8, 13, and 21. Clearly, the Fibonacci numbers have emerged again.

Using the Fibonacci Numbers to Convert Kilometers to and from Miles

Perhaps the most common measure of distance today is the kilometer. Yet the United States still uses the mile to measure distance. Converting from

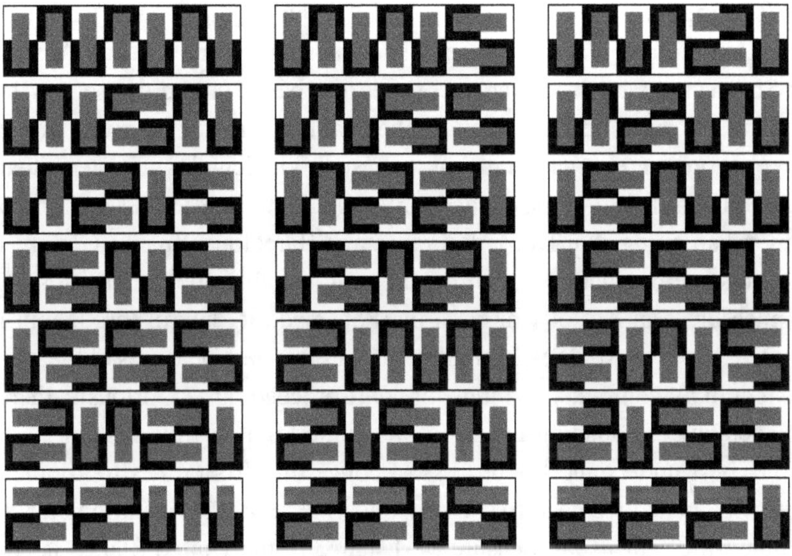

Figure 2-9

one unit of measure to the other can be done with the help of the Fibonacci numbers. As soon as one becomes familiar with these Fibonacci numbers, the conversions can be done mentally and almost immediately. Before we discuss the conversion process between these two units of measure, let's look at their origin.

The mile derives its name from the Latin word for 1,000, *mille*, as it represented the distance that a Roman legion could march in 1,000 paces (which is 2,000 steps). One of these paces was about 5 feet, so the Roman mile was about 5,000 feet. The Romans marked off these miles with stones along the many roads they built in Europe—hence the name, "milestones!" The term "statute mile" goes back to Queen Elizabeth I of England, who redefined the mile from 5000 feet to 8 furlongs[1] (5280 feet) by statute in 1593.

[1]A **furlong** is a measure of distance within Imperial units and U.S. customary units. Although its definition has varied historically, in modern terms it equals 660 feet, and is therefore equal to 201.168 meters. There are eight furlongs in a mile. The name "furlong" derives from the Old English words *furh* (furrow) and *lang* (long). It originally referred to the length of the furrow in one acre of a ploughed open field (a medieval communal field which was divided into strips). The term is used today for distances horses run at a racetrack.

The metric system dates back to 1790 during the French Revolution, when the French National Assembly requested that the French Academy of Sciences establish a standard of measure based on the decimal system, which they did. The unit of length they called a "meter," derived from the Greek word *metron*, which means "measure." Its length was determined to be one 10 millionth of the distance from the North Pole to the equator along a meridian going near Dunkirk, France and Barcelona, Spain.[2] Clearly, the metric system is better suited for scientific use than the American system of measure. An Act of Congress in 1866 made it "lawful throughout the United States of America to employ the weights and measures of the metric system in all contracts, dealings, or court proceedings." Curiously, there is no such law establishing the use of the United States' mile system.

To best understand the process of converting from miles to kilometers (and vice versa) we need to see how one mile relates to one kilometer. The statute mile (our usual measure of distance in the United States) is equal to 1.609344 kilometers. One the other hand, one kilometer has a length of 0.621371192 miles. The nature of these two numbers, where we have reciprocals that differ by about 1, might remind us of the Golden Ratio, which is approximately 1.618, and its reciprocal is approximately 0.618. This would tell us that the Fibonacci numbers, the ratio of whose consecutive members approaches the Golden Ratio, might come into play here. For starters, we will consider a length of 5 miles, which, when converted to kilometers, would be $5 \times 1.609344 = 8.04672 \approx 8$.

We could also check to see what the equivalent of 8 kilometers would be in miles: $8 \times .621371192 = 4.970969536 \approx 5$. This allows us to conclude that approximately 5 miles is about equal to 8 kilometers—two consecutive Fibonacci numbers. Therefore, since the relationship between miles and kilometers is very close to the Golden Ratio, they appear to be almost in the relationship of consecutive Fibonacci numbers. Using this relationship to *approximately* convert 13 kilometers to miles, replace 13 with the previous Fibonacci number, 8, and we can conclude that 13 kilometers is

[2]There are three fascinating books on this subject: Dava Sobel, *Longitude*, (New York: Walker & Co., 1995); Umberto Eco, *The Island of the Day Before*, (New York: Harcourt Brace, 1994); and Thomas Pynchon, *Mason & Dixon*, (New York: Henry Holt, 1997).

about equal to 8 miles. Similarly, 5 kilometers is about equal to 3 miles and 2 kilometers is about equal to 1 mile. Larger Fibonacci numbers will give us a more accurate estimate, since the ratio of these larger consecutive numbers gets closer to the Golden Ratio, φ, which will be presented in the next chapter.

To become more familiar with this procedure, suppose you want to convert 20 kilometers to miles. We have selected 20, since it is *not* a Fibonacci number. We can express 20 as a sum of Fibonacci numbers and convert each number separately and then add them. Thus, 20 kilometers = 13 kilometers + 5 kilometers + 2 kilometers, which, by replacing 13 by 8, and 5 by 3, and 2 by 1, is approximately equal to 12 miles. Another way of doing that would be to consider the 20 kilometers as 4×5 kilometers. Since 5 kilometers is equal to approximately 3 miles, we would simply multiply $3 \times 4 = 12$ miles.

To use this process to achieve the reverse, that is, to convert miles to kilometers, we write the number of miles as a sum of Fibonacci numbers and then replace each by the next *larger* Fibonacci number. Converting 20 miles to kilometers: 20 miles = 13 miles + 5 miles + 2 miles. Now, replacing each of the Fibonacci numbers with the next larger in the sequence, we get: 20 miles = 21 kilometers + 8 kilometers + 3 kilometers = 32 kilometers.

When finding the Fibonacci representation of a number, there is no need to use the fewest Fibonacci numbers. You can use any combination of numbers whose sum is the number you are converting. For instance, 40 kilometers is 2×20 and we have just seen that 20 kilometers is 12 miles. So, 40 kilometers is $2 \times 12 = 24$ miles (approximately).

It is not a trivial matter for us to conclude that every natural number can be expressed as the sum of other Fibonacci numbers without repeating any one of them in the sum. Let's take the first few Fibonacci numbers to demonstrate this property. You should begin to see patterns, and note that we used the fewest number of Fibonacci numbers in each sum in Figure 2-10.

For example, we could also have represented 21 as the sum of $3 + 5 + 13$, or as $8 + 13$. It could be an entertaining challenge to represent given numbers in terms of the fewest Fibonacci numbers necessary. The Belgian mathematician Edouard Zeckendorf (1901–1983) proved that each natural number can be expressed as a unique sum of nonconsecutive Fibonacci numbers.

n	**The sum of Fibonacci numbers equal to *n***
1	1
2	2
3	3
4	1+3
5	5
6	1+5
7	2+5
8	8
9	1+8
10	2+8
11	3+8
12	1+3+8
13	13
14	1+13
15	2+13
16	3+13

Figure 2-10 Numbers Expressed as Sums of Fibonacci Numbers

Fibonacci Numbers in Physics

A problem posed in 1963 in the field of optics shows a surprise occurrence of the Fibonacci numbers.[3] The problem begins with two glass plates placed face to face as shown in Figure 2-11.

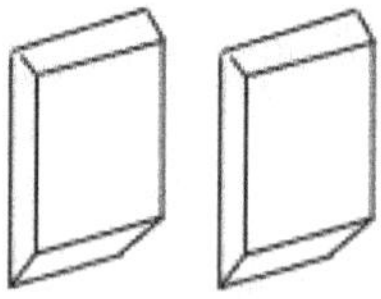

Figure 2-11

[3]This problem was posed by L. Moser and M. Wyman, "Problem B-6," *The Fibonacci Quarterly*, 1:1 (Feb. 1963), 74, and solved by L. Moser and J.L. Brown, "Some Reflections," *The Fibonacci Quarterly*, 1:1 (Dec. 1963), 75.

The problem is to determine the number of possible reflections when intermittently covering a surface at the back side. For convenience, we shall number the various surfaces of reflection as in Figure 2-12.

The first case, where there are no reflections, is where the light source passes right through both plates of glass, as shown in Figure 2-13. This is the **one** path the light ray takes.

The next case is where there is one reflection. There can be **two** possible paths that the light can take in this case. It can reflect off the two surfaces shown in Figure 2-14.

Suppose that the ray of light is reflected twice within this set of two plates of glass. Then there will be **three** possible reflections, as can be seen in Figure 2-15.

If the light ray is reflected three times, then there will be **five** possible paths that the ray can take, as seen in Figure 2-16.

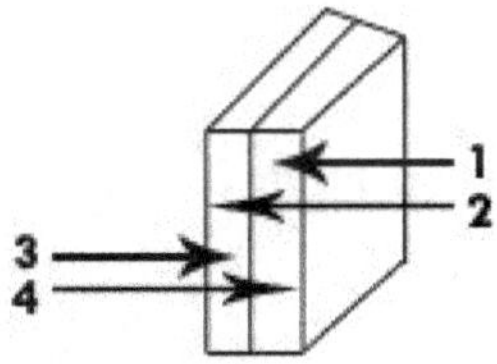

Figure 2-12

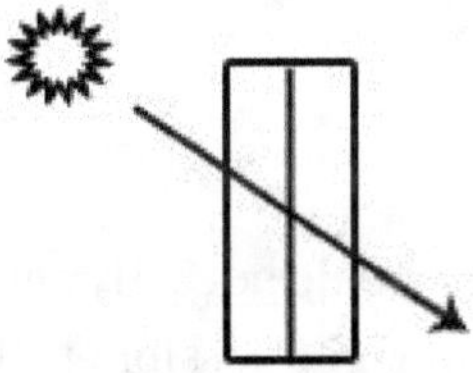

Figure 2-13

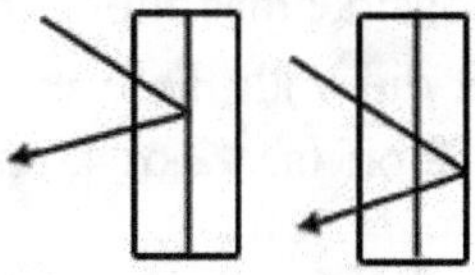

Figure 2-14

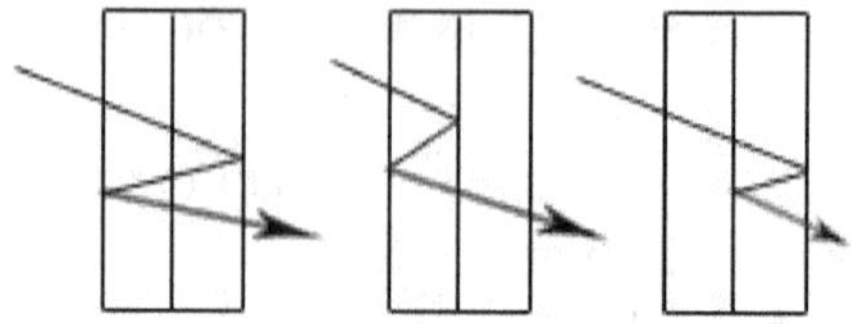

Figure 2-15

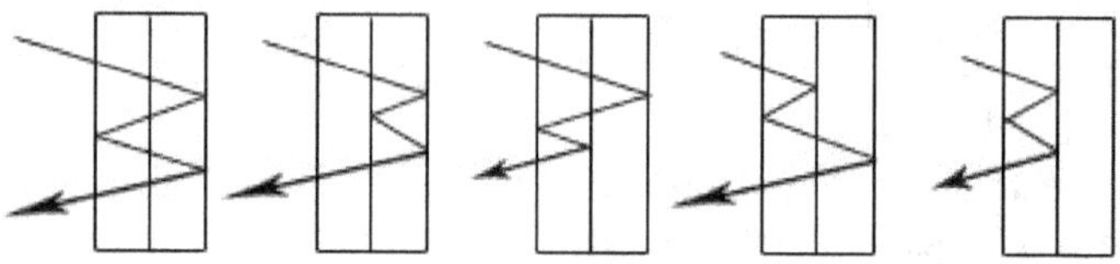

Figure 2-16

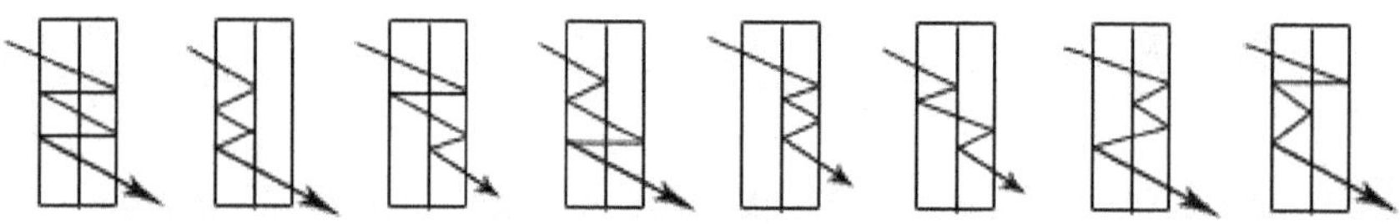

Figure 2-17

Figure 2-18

At this point, the next case could be easily anticipated: the light ray taking four reflections will result in **eight** possible paths, as seen in Figure 2-17.

The pattern evolving should be clear by now. The Fibonacci numbers are represented consecutively by the number paths shown in Figures 2-13–2-17. To generalize this pattern, consider the following. Suppose the last reflection occurs at face 1 or 3, as it did with the even number of reflections. The previous reflection had to be off face 2 or 4, respectively, as demonstrated in Figure 2-18.

If the last of n reflections was off face 1, then there were $(n-1)$ reflections prior to this one, or $(n-1)$ possible paths, shown in Figure 2-18. If the last reflection (of n reflections) was off face 3, then the previous reflection—the $(n-1)^{\text{th}}$ reflection—must have occurred off face 4, and must have had $(n-2)$ reflections prior to this one. And so there were $(n-2)$ possible paths. Therefore, since both reflections off face 1 or 3 can be considered the last reflection, we get the number of paths as $(n-1)+(n-2)$, which would indicate that F_n is the number of paths, the sum of the 2 previous numbers. This began with zero reflections giving us **1** path, then one reflection, which gave us **2** paths, and three reflections, which gave us **3** paths. This implies that F_{n+2} indicates the number of paths for n reflections. Again, the Fibonacci numbers appear!

Displaying a Watch

When one least expects it, the Golden Ratio appears. For a variety of reasons, whenever you see an advertisement for a clock, the face, more often than not, displays a time approximately 10:10, as shown in Figure 2-19.

At 10:10, the hands of a clock form an angle of $19\frac{1}{6}$ minutes,[4] which is the equivalent of 115 degrees (Figure 2-19). Now consider the rectangle formed

Figure 2-19 A Golden Angle

[4]Since 10 minutes is $\frac{1}{6}$ of an hour, the hour hand moved $\frac{1}{6}$ of the distance from "10" to "11" from 10:00 to 10:10. Therefore, the hour hand moved $\frac{1}{6}$ of 5 minute markers, or $\frac{5}{6}$ of a minute

Figure 2-20 Golden Angle

by placing its vertices at points at which the hands indicate 10:10, that is, where the minute marker points to 2 and the hour marker points to $10\frac{1}{6}$. You can see this rectangle in Figure 2-20, where the point of intersection of its diagonals are at the center of the watch face. This rectangle is very close to the Golden Rectangle, whose diagonals meet at an angle of about 116.6 degrees, which can be seen in Figure 2-21 (and discussed in greater detail in the next chapter).

We use the Golden Rectangle and find the measure of the angle where $\tan x = \frac{FB}{FE} = \frac{\frac{\phi}{2}}{\frac{1}{2}} = \phi$. That angle, x, measures 58.28 degrees. Therefore, the angles formed by the diagonals, namely, $m\angle AEB = 116.56^\circ$, which is very close to 115°. Whether consciously or subconsciously, in almost all advertisements showing a watch face, the hands of the clock are pointed approximately at 10:10 o'clock. As expected, this is not always the case. As we can see in Figure 2-22, for example, a United States postage stamp exhibits the "American clock," which reads about $8{:}21\frac{1}{2}$.

---

marker. Thus, the angle at 10:10 is $19\frac{1}{6}$ minute markers. To find the degree measure of that angle, we simply find the part of 60-minute markers that $19\frac{1}{6}$ is: $\frac{19\frac{1}{6}}{60} = \frac{\frac{115}{6}}{60} = \frac{115}{360} = 115^\circ$.

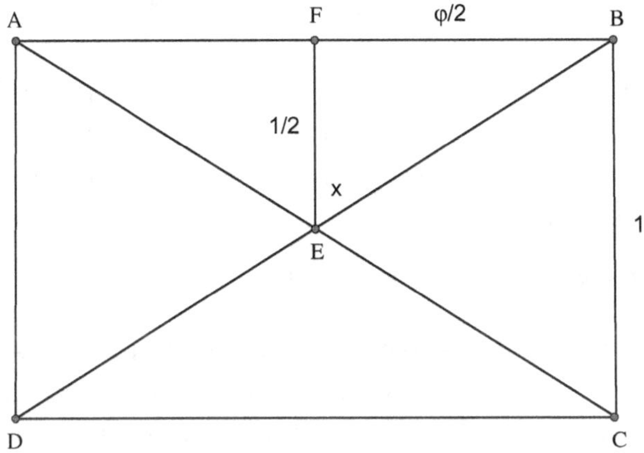

Figure 2-21

Figure 2-22 Golden Angle

At 8:21$\frac{1}{2}$, the hands form an angle of about 112.79167 degrees,[5] which is not exactly the angle of the diagonals of an ideal Golden Rectangle. But

[5] 60 min = 360°, 1 min = 6°, 10 min = 60°. Therefore, $\angle(12; 8{:}21\frac{1}{2}) = 21\frac{1}{2} \cdot 6° = 129°$, and $\angle(6; 21\frac{1}{2}) = 180° - 129° = 51°$. The hour hand's location: 12 h = 360°, 1 h = 30°, 10 min = $\frac{1}{6}$h = 5°, 1 min = $\frac{1}{60}$h = $\frac{1}{12}$°, $\angle(12; 8{:}00) = 8 \cdot 30° = 240°$, $\angle(8{:}00; 8{:}21\frac{1}{2}) = 21{,}5 \cdot \frac{1}{12}° = 1.79167°$. Therefore, $\angle(12; 8{:}21\frac{1}{2}) = 241.79167°$, and $\angle(6; 8{:}21\frac{1}{2}) = 241.79167° - 180° = 61.79167°$. The sum = 51° + 61.79167° = **112.79167°**.

it is close enough for the naked eye to appreciate. Once again, the Golden Rectangle seems to have captured human beauty.

When taught to drive a car, learning drivers are often instructed to hold the steering wheel at the positions of 10 and 2 on a clock. Perhaps those points present some sort of special balance. Something to ponder!

Fibonacci Helps Us Determine Seatings

Curiously, when one least expects it, the Fibonacci numbers once again appear to help us explain a problem, which turns into a pattern using this famous sequence. There are interesting mathematical arrangements when setting up specified seating. The problem here involves a school full of boys and girls that need to be seated on the stage so that no two boys sit next to each other. The chart in Figure 2-23 summarizes the various seating arrangements, where the B stands for boy and the G stands for girl.

Number of chairs	**Arrangements**	**Number of ways to seat students**
1	B, G	2
2	BG, GB, GG, ~~BB~~	3
3	GGB, ~~BBG~~, BGB, GBG, GGG, ~~BBB~~, BGG, ~~GBB~~,	5
4	GGGG, ~~BBBB~~, ~~BBGG~~, ~~GGBB~~, BGBG, GBGB, ~~GBBG,~~ BGGB, ~~BBBG~~, GGGB, ~~GBBB~~, BGGG, ~~BBGB~~, GGBG, ~~BGBB~~, GBGG	8
5	GGGGB, ~~BBBBB~~, ~~BBGGB~~, ~~GGBBB~~, BGBGB, ~~GBGBB~~, ~~GBBGB,~~ ~~BGGBB~~, ~~BBBGB~~, ~~GGGBB~~, ~~GBBBB~~, BGGGB, ~~BBGBB~~, GGBGB, ~~BGBBB~~, GBGGB GGGGG, ~~BBBBG~~, ~~BBGGG~~, ~~GGBBG~~, BGBGG, GBGBG, ~~GBBGG,~~ BGGBG, ~~BBBGG~~, GGGBG, ~~GBBBG~~, BGGGG, ~~BBGBG~~, GGBGG, ~~BGBBG~~, GBGGG	13

Figure 2-23 Seating Arrangements

As you can clearly see, in each case the number of seating possibilities is a successive Fibonacci number, beginning with 2.

Let us now change the restrictions so that no boy or girl sits without someone from the same sex next to him/her *and* that the first position must be a girl. Figure 2-24 displays the various acceptable arrangements, once again showing that the Fibonacci numbers indicate the number of ways this seating can be done.

Number of chairs	Arrangements	Number of ways to seat students
1	~~B, G~~	0
2	~~BG, GB~~, GG, ~~BB~~	1
3	~~GGB~~, ~~BBG~~, ~~BGB~~, ~~GBG~~, GGG, ~~BBB~~, ~~BGG~~, ~~GBB~~,	1
4	GGGG, BBBB, ~~BBGG~~, CCDD, ~~BGBG~~, ~~GBGB~~, ~~GBBG~~, ~~BGGB~~, ~~BBBG~~, ~~GGGB~~, ~~GBBB~~, ~~BGGG~~, ~~BBGB~~, ~~GGBG~~, ~~BGBB~~, ~~GBGG~~	2
5	~~GGGGB~~, ~~BBBBB~~, ~~BBGGB~~, GGBBB, ~~BGBGB~~, ~~GBGBB~~, ~~GBBGB~~, ~~BGGBB~~, ~~BBBGB~~, GGGBB, ~~GBBBB~~, ~~BGGGB~~, ~~BBGBB~~, ~~GGBGB~~, ~~BGBBB~~, ~~GBGGB~~ GGGGG, ~~BBBBG~~, ~~BBGGG~~, ~~GGBBG~~, ~~BGBGG~~, ~~GBGBG~~, ~~GBBGG~~, ~~BGGBG~~, ~~BBBGG~~, ~~GGGBG~~, ~~GBBBG~~, ~~BGGGG~~, ~~BBGBG~~, ~~GGBGG~~, ~~BGBBG~~, ~~GBGGG~~	3

Figure 2-24 Seating Arrangements

Fish in a Hatchery

We begin with a situation that may not be too realistic, but it provides a nice application of the Fibonacci numbers. Imagine a fish hatchery that is partitioned into 16 congruent regular hexagons arranged in two rows, as shown in Figure 2-25, with pathways between adjacent hexagons. A fish begins a journey at the upper left side of the double row of hexagons and

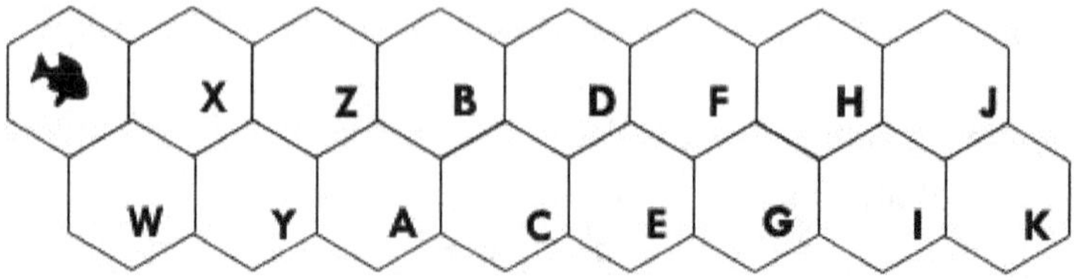

Figure 2-25

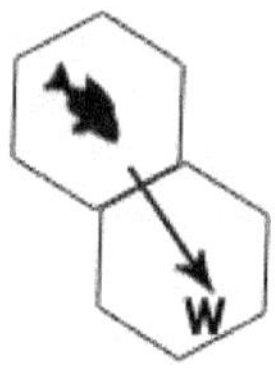

Figure 2-26

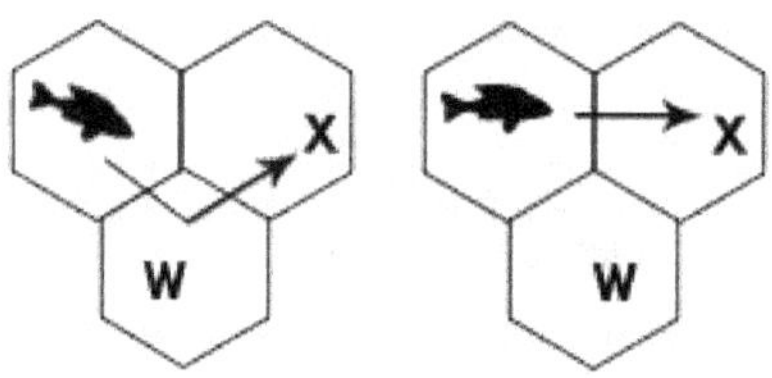

Figure 2-27

ends up at the lower right hexagon. Our problem is to determine the number of paths the fish can take to complete his journey to reach hexagon K, if he can only move to the right.

As the fish begins traveling, there is only *one* path that can be taken to hexagon W, since if he were to go horizontally first, he would have to go left to get to hexagon W (see Figure 2-26).

To get to hexagon X, there are *two* paths, one directly to the right, and one through W (see Figure 2-27).

For the fish to get to hexagon Y, there are *three* paths: hexagons W-X-Y, X-Y, and W-Y, as can be seen in Figure 2-28.

By now we can anticipate the next number of possible hexagon-paths that the fish can travel when there are five hexagons available. Here there are *five* paths, as shown in Figure 2-29. They are: W-Y-Z, W-X-Y-Z, X-Z, W-X-Z, and X-Y-Z.

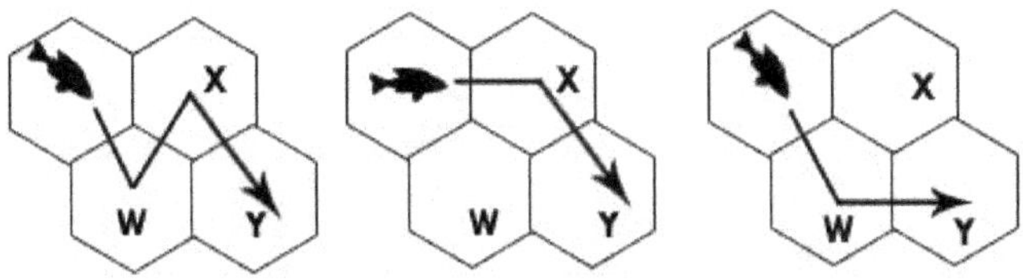

Figure 2-28

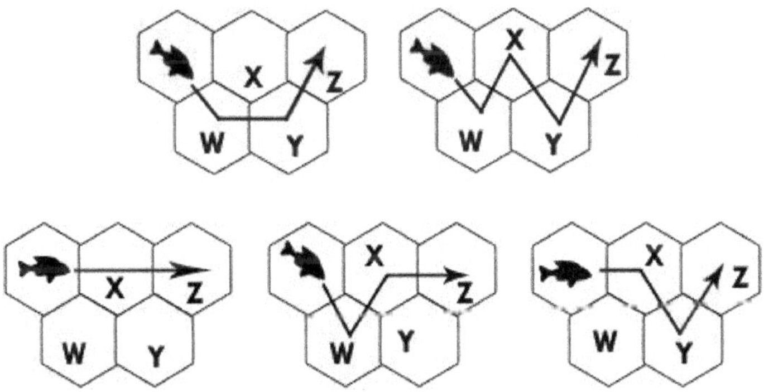

Figure 2-29

Number of hexagons for the fish's journey	1	2	3	4	5	6	7	...	n
Number of different paths	1	1	2	3	5	8	13		F_n

Figure 2-30 Summary of Fish Paths Through Hexagons

It would be correct to assume that this pattern will continue as we increase the number of hexagons for the fish to traverse. A summary of these paths is provided in Figure 2-30.

So, we might conclude that for the fish to reach the 16th hexagon, K, we would need to find the 16th Fibonacci number, F_{16}, which is 987. Thus, there are 987 possible paths for the fish to travel from the first hexagon to the 16th hexagon, K, always traveling only to the right. Now that we have learned about the Fibonacci numbers and the many places that they tend to appear, we are ready to embark on a more physical aspect of these numbers. Curiously, the Fibonacci numbers are very closely related to the Golden Ratio: as the numbers in the Fibonacci sequence get larger, the ratio of the consecutive pairs approaches the Golden Ratio. The next chapter will take us through a geometric appreciation of this amazing ratio.

Chapter 3

Introduction to the Golden Ratio

There are many aspects of mathematics that escape common knowledge. The general public tends to dislike the subject, and these aspects need to be exhibited and explained to change that negative attitude. One such is the Golden Ratio, which got the name because of its prevalence as a design element and its seemingly universal aesthetic appeal. The famous German mathematician Johannes Kepler (1571–1630) described this ratio's importance by saying that "geometry harbors two great treasures: one is the Pythagorean theorem, and the other is the Golden Ratio. The first can compare with a heap of gold and the second we simply call a priceless jewel."

The Golden Ratio manifests itself in countless everyday appearances. The human body displays the Golden Ratio, as do the helix structure of DNA, the design of Greek architecture, and many modern masterpieces. The Golden Ratio is a unique pattern that has not only wide-ranging applications but also manifests itself in the most unexpected places. To appreciate where the Golden Ratio appears, we first need to understand exactly what it is. This will allow us to recognize it and truly appreciate it when it reveals itself.

Defining the Golden Ratio

The Golden Ratio is the ratio of the lengths of the two parts of a line segment in the following proportion: the longer segment (L) is to the shorter

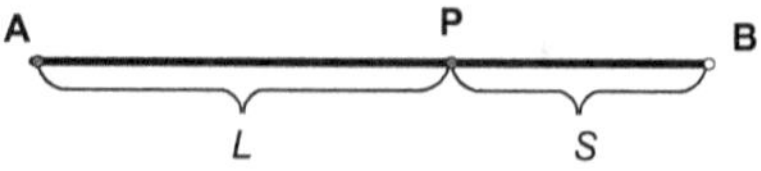

Figure 3-1 Golden Ratio

segment (S) as the entire original segment ($L + S$) is to the longer segment (L). Symbolically, this is written as $\frac{L}{S} = \frac{L+S}{L}$, as shown in Figure 3-1.

This is called the *Golden Ratio*, and is sometimes referred to as the *Golden Section*; the latter case refers to the "sectioning" or partitioning of a line segment. The terms Golden Ratio and Golden Section were first introduced by the Franciscan friar and mathematician Fra Luca Pacioli (ca. 1445–1514 or 1517), who was the first to use the term as the title of his 1509 book *De divina proportione* (Divine Proportion). The German mathematician Martin Ohm (1792–1872) was the first to use the term *Goldener Schnitt* (Golden Section). In English, this term *Golden Section* was first used by James Sully in 1875 in an article on aesthetics in the 9th edition of the Encyclopedia Britannica.

To determine the numerical value of the Golden Ratio $\frac{L}{S}$ we will change this equation $\frac{L}{S} = \frac{L+S}{L}$, or $\frac{L}{S} = \frac{L}{L} + \frac{S}{L}$, to an equivalent, where $x = \frac{L}{S}$, to get $x = 1 + \frac{1}{x}$. This can be rewritten as $x^2 - x - 1 = 0$. We can now solve this equation for x by using the quadratic formula to obtain the numerical value of the Golden Ratio: $\frac{L}{S} = x = \frac{1+\sqrt{5}}{2}$, which is commonly denoted by the Greek letter phi: ϕ.[1] Thus, $\phi = \frac{L}{S} = \frac{1+\sqrt{5}}{2} \approx 1.61803398874989485$. What makes this ratio $\frac{L}{S}$ mathematically unique is its reciprocal, namely, $\frac{S}{L} = \frac{1}{\phi} = \frac{2}{1+\sqrt{5}} \approx 0.61803398874989485$. Now, you should notice a very unusual relationship. The value of ϕ and $\frac{1}{\phi}$ differ by 1. That is, $\phi - \frac{1}{\phi} = 1$. From the normal relationship of reciprocals, the product of ϕ and $\frac{1}{\phi}$ is also

[1]There is reason to believe that the letter ϕ was used because it is the first letter of the name of the celebrated Greek sculptor Phidias (ca. 490–430 BCE) [in Greek: ΦΕΙΔΙΑΣ or $\Phi\varepsilon\iota\delta\iota\alpha\varsigma$ (Pheidias)], who produced the famous statue of Zeus in the Temple of Olympia and supervised the construction of the Parthenon in Athens, Greece. Phidias's frequent use of the Golden Ratio in this glorious building is likely the reason for this attribution, although there is no direct evidence that Phidias consciously used this ratio.

The American mathematician Mark Barr was the first to use the letter ϕ in about 1909. See Theodore Andrea Cook, *The Curves of Life*, (New York: Courier Dover Publications, 1914, reprinted 1979) p. 420.

equal to 1, that is, $\phi \cdot \frac{1}{\phi} = 1$. Therefore, we have two numbers, ϕ and $\frac{1}{\phi}$, whose difference *and* product are both equal to 1. These are the only two numbers for which this is true! By the way, you might have noticed that the sum of these two numbers is $\phi + \frac{1}{\phi} = \sqrt{5}$, since $\frac{\sqrt{5}+1}{2} + \frac{\sqrt{5}-1}{2} = \sqrt{5}$.

Constructing the Golden Ratio

The challenge here is to partition a line segment into the Golden Ratio. We will use the term Golden Ratio to refer to the numerical value of ϕ, and the term Golden Section to refer to the geometric division of a segment into the ratio ϕ.

Golden Ratio Construction 1

Perhaps the most popular way of constructing the Golden Section is to begin with a unit square[2] $ABCD$, with midpoint M of side AB, and then draw a circular arc MC, cutting the extension of side AB at point E. We now can claim that line segment AE is partitioned into the Golden Section at point B.

To prove this, we will apply the definition of the Golden Section $\frac{AB}{BE} = \frac{AE}{AB}$ to see whether it holds true. Substituting the values obtained by applying the Pythagorean theorem to ΔMBC as shown in Figure 3-2, we get the

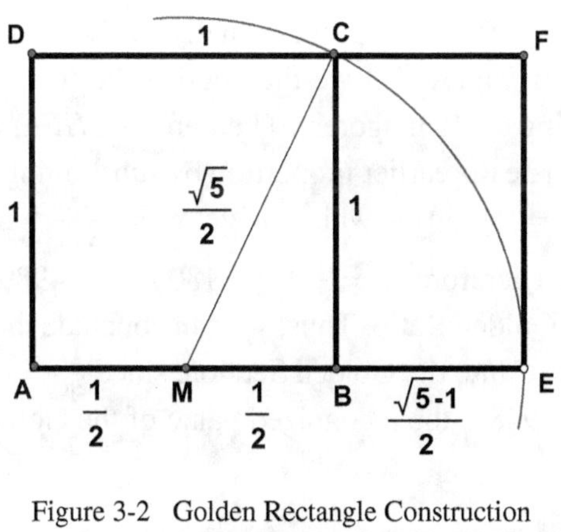

Figure 3-2 Golden Rectangle Construction

[2]A *unit square* is a square with side length of 1 unit.

following: $MC^2 = MB^2 + BC^2 = \left(\frac{1}{2}\right)^2 + 1^2 = \frac{1}{4} + 1 = \frac{5}{4}$; therefore, $MC = \frac{\sqrt{5}}{2}$. It follows that $BE = ME - MB = MC - MB = \frac{\sqrt{5}}{2} - \frac{1}{2} = \frac{\sqrt{5}-1}{2}$ and $AE = AB + BE = 1 + \frac{\sqrt{5}-1}{2} = \frac{2}{2} + \frac{\sqrt{5}-1}{2} = \frac{\sqrt{5}+1}{2}$. We then can find the value of $\frac{AB}{BE} = \frac{AE}{AB}$, that is, $\frac{1}{\frac{\sqrt{5}-1}{2}} = \frac{\frac{\sqrt{5}+1}{2}}{1}$, which turns out to be a true proportion since the cross products are equal. That is, $(\frac{\sqrt{5}-1}{2}) \cdot (\frac{\sqrt{5}+1}{2}) = 1 \cdot 1 = 1$.

We can also see from Figure 3-2 that point B can divide the line segment AE into an *inner* Golden Section, since $\frac{AB}{AE} = \frac{1}{\frac{\sqrt{5}-1}{2}} = \frac{\sqrt{5}+1}{2} = \phi$. Meanwhile, point E can divide the line segment AB into an *outer* Golden Section, since $\frac{AE}{AB} = \frac{1+\frac{\sqrt{5}-1}{2}}{1} = \frac{\sqrt{5}+1}{2} = \phi$. Notice the shape of the rectangle $AEFD$ in Figure 3-2. The ratio of the length to the width is the Golden Ratio: $\frac{AE}{EF} = \frac{\frac{\sqrt{5}+1}{2}}{1} = \frac{\sqrt{5}+1}{2} = \phi$. This appealing shape is called the *Golden Rectangle*, and plays a dominant role in architecture, art, and beyond.

Golden Ratio Construction 2

There are many other ways to construct the Golden Section. Here we begin with the three adjacent unit squares shown in Figure 3-3. We construct the angle bisector of $\angle BHE$. There is a convenient geometric relationship that will be very helpful to us here; that is, the angle bisector in a triangle divides the side to which it is drawn proportionally to the two legs of the angle being bisected.[3] In Figure 3-3 we then derive the following relationship: $\frac{BH}{EH} = \frac{BC}{CE}$. Applying the Pythagorean Theorem to ΔHFE, we get $HE = \sqrt{5}$. We can now evaluate the earlier proportion by substituting the values shown in Figure 3-3: $\frac{1}{\sqrt{5}} = \frac{x}{2-x}$, from which we get $x = \frac{2}{\sqrt{5}+1}$, which is the reciprocal of $\frac{\sqrt{5}+1}{2} = \phi$. Therefore, $x = \frac{1}{\phi} \approx 0.61803398874989485$, which is the reciprocal of the Golden Ratio. Thus, we can conclude that point B divides the line segment AC into the Golden Section, since $\frac{AB}{BC} = \frac{1}{x} = \frac{\sqrt{5}+1}{2} = \phi \approx 1.61803398874989485$, the recognized value of the Golden Ratio.

[3]This theorem was originally demonstrated by Euclid in Book VI, § 3 of his *Elements*. The proof can also be found in A. S. Posamentier and R. Bannister, *Geometry: Its Elements and Structure*, 2nd Edition, (New York: Dover, 2014).

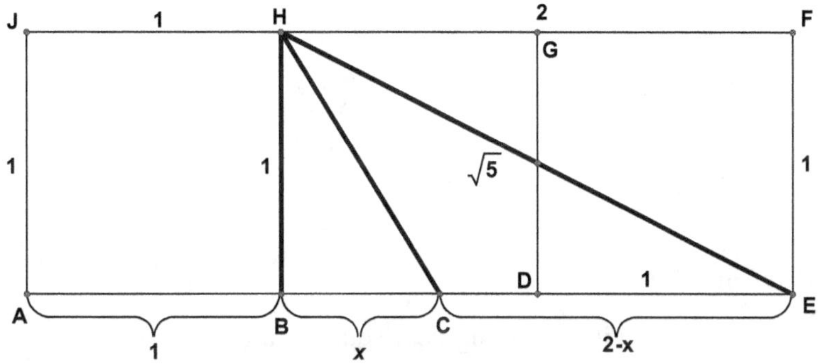

Figure 3-3 Golden Rectangle Construction

Golden Ratio Construction 3

Another method for constructing the Golden Section is credited to Heron of Alexandria (ca. 40–120) and begins with the construction of a right triangle with one leg of unit length and the other leg twice as long, as shown in Figure 3-4. Here we will partition the line segment *AB* into the Golden Ratio.

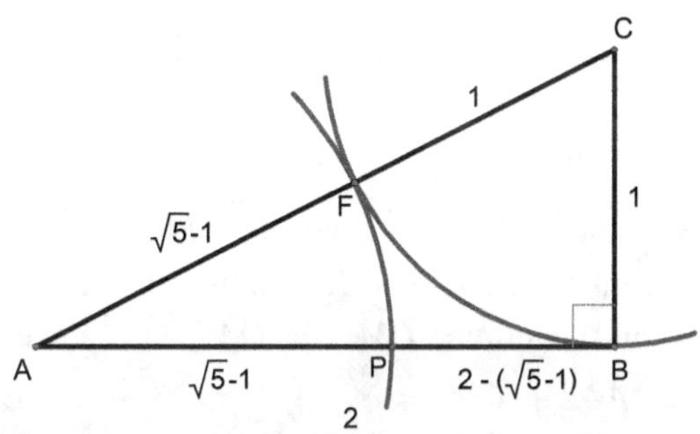

Figure 3-4 Golden Ratio Construction

With $AB = 2$ and $BC = 1$, we apply the Pythagorean theorem to ΔABC. We then find that $AC = \sqrt{2^2 + 1^2} = \sqrt{5}$. With the center at point C, we draw a circular arc with radius 1, cutting line segment AC at point F. Then we draw a circular are with the center at point A and the radius AF, cutting AB at point P. Because $AF = \sqrt{5} - 1$, we get $AP = \sqrt{5} - 1$. Therefore, $BP = 2 - (\sqrt{5} - 1) = 3 - \sqrt{5}$. To determine the ratio $\frac{AP}{BP}$, we will set up the ratio $\frac{\sqrt{5}-1}{3-\sqrt{5}}$ and then to make this more manageable, we will rationalize the denominator by multiplying the ratio by 1 in the form of $\frac{3+\sqrt{5}}{3+\sqrt{5}}$. We then find that $\frac{\sqrt{5}-1}{3-\sqrt{5}} \cdot \frac{3+\sqrt{5}}{3+\sqrt{5}} = \frac{3\sqrt{5}+5-3-\sqrt{5}}{3^2-(\sqrt{5})^2} = \frac{2\sqrt{5}+2}{9-5} = \frac{2(\sqrt{5}+1)}{4} = \frac{\sqrt{5}+1}{2} = \phi \approx$ 1.618033988, which is the Golden Ratio! Therefore, point P cuts the line segment AB into the Golden Ratio.

Golden Ratio Construction 4

This proof begins with two congruent squares, as shown in Figure 3-5. A circle is drawn with its center at the midpoint M of the common side of the squares, and its radius is half the length of the side of the square. The intersection of the circle and the diagonal of the rectangle at C determines the Golden Section AC with respect to a side of the square, AD.

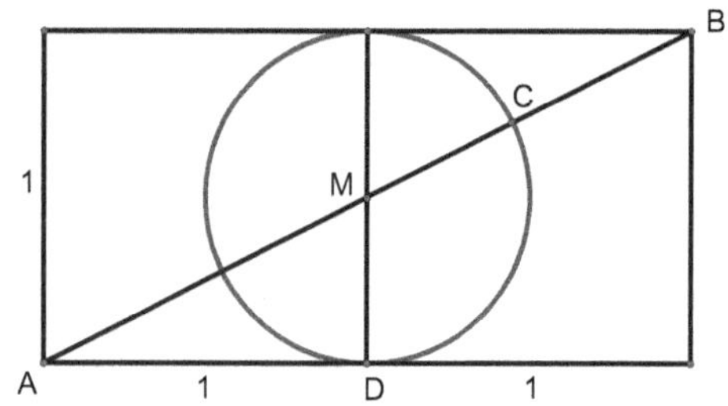

Figure 3-5 Golden Ratio Construction

With $AD = 1$ and $DM = \frac{1}{2}$, we get $AM = \frac{\sqrt{5}}{2}$ by applying the Pythagorean theorem to triangle AMD, shown in Figure 3-6. Since CM is also a radius of the circle, $CM = DM = \frac{1}{2}$. We can then conclude that $AC = AM + CM = \frac{\sqrt{5}}{2} + \frac{1}{2} = \frac{\sqrt{5}+1}{2} = \phi$. Furthermore, $BC = AB - AC = \sqrt{5} - \frac{\sqrt{5}+1}{2} = \frac{\sqrt{5}-1}{2} = \frac{1}{\phi}$.

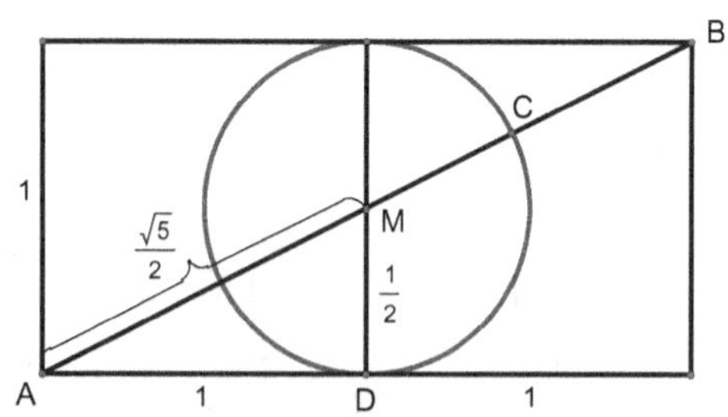

Figure 3-6 Golden Ratio Construction

Golden Ratio Construction 5

Here we will show that the semicircle on the extended side of a square, whose radius is the distance from the midpoint of the side of the square to an opposite vertex, creates a line segment where the vertex of the square determines the Golden Ratio. In Figure 3-7, we have square $ABCD$ and a semicircle on line AB, with center at the midpoint M of AB, and radius CM. We encountered a similar situation with Construction 1, where we concluded that $\frac{AB}{BE} = \phi$, and $\frac{AE}{AB} = \phi$.

However, here we have an extra added attraction: DE and BC partition each other into the Golden Section at point P. This is easily justified in that triangles DPC and EBP are similar and their corresponding sides DC and BE are in the Golden Ratio. Hence, all the corresponding sides are in the Golden Ratio, which here is $\frac{CP}{PB} = \frac{DP}{PE} = \phi$.

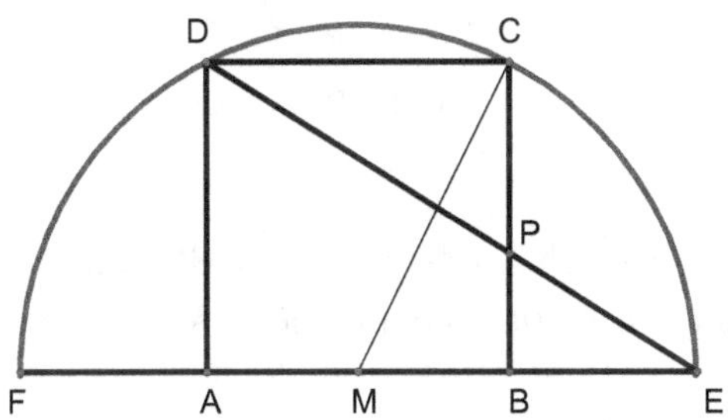

Figure 3-7 Golden Ratio Construction

Golden Ratio Construction 6

For this construction we will consider the inscribed equilateral triangle ABC with two sides bisected by line segment PT at points Q and S, as shown in Figure 3-8. We begin by letting the side length of the equilateral triangle equal 2, which supplies us with the segment lengths indicated in Figure 3-8. The proportionality there gives us $\frac{RS}{CD} = \frac{AS}{AC}$, and substituting appropriate values yields $\frac{RS}{1} = \frac{1}{2}$ and $RS = \frac{1}{2}$.

Due to the symmetry of the figure, a useful geometric theorem will enable us to find the length of the segments $PQ = ST = x$. The theorem states that the products of the segments of two intersecting chords of a circle are equal. From that theorem, we find $PS \cdot ST = AS \cdot SC$, from which we

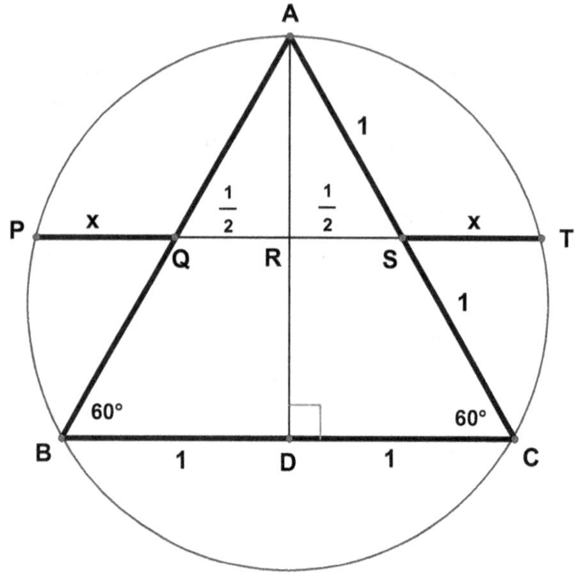

Figure 3-8 Golden Ratio Construction

get $(x+1)x = 1 \cdot 1$, which is $x^2 + x - 1 = 0$, leading to $x = \frac{\sqrt{5}-1}{2}$. Therefore, the segment QT is partitioned into the Golden Section at point S, since $\frac{QS}{ST} = \frac{1}{x} = \frac{2}{\sqrt{5}-1} = \frac{\sqrt{5}+1}{2} \approx 1.61803398874989485$, which we recognize as the value of the Golden Ratio. We can generalize this construction by saying that the midline of an equilateral triangle extended to the circumcircle is partitioned into the Golden Section by the sides of the equilateral triangle.

Golden Ratio Construction 7

This is a rather easy construction of the Golden Ratio that simply requires constructing an isosceles triangle inside a square as shown in Figure 3-9. The vertex E of ΔABE lies on side DC of square $ABCD$, and altitude EM intersects the inscribed circle of ΔABE at point H. The Golden Ratio appears in two ways here. When the side of the square is 2, then the radius of the inscribed circle $r = \frac{1}{\phi}$, and the point H partitions EM into the Golden Ratio as $\frac{EM}{HM} = \phi$.

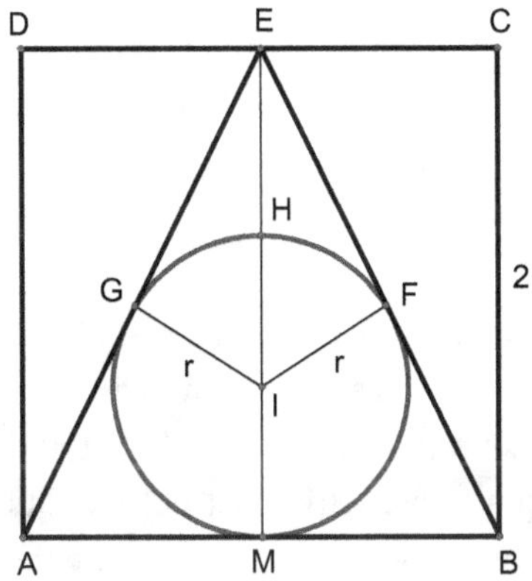

Figure 3-9 Golden Ratio Construction

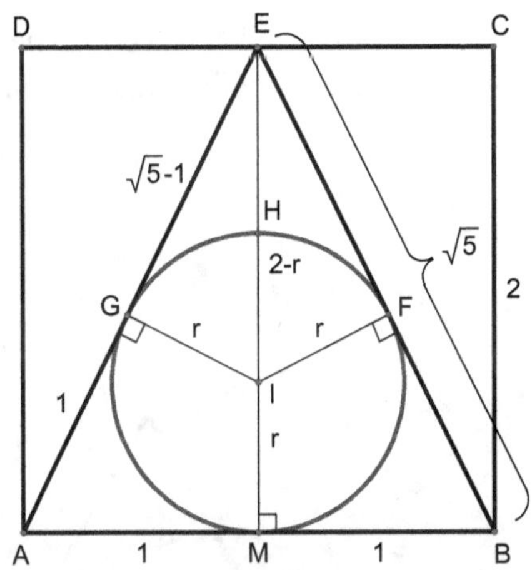

Figure 3-10 Golden Ratio Construction

To justify this construction, we will let the side of the square have length 2. This gives us $BM = 1$ and $EM = 2$. Then, with the Pythagorean theorem applied to triangle *MEB*, we derive $AE = BE = \sqrt{5}$, whereupon we recognize that $GE = \sqrt{5} - 1$, which is shown in Figure 3-10. We apply the Pythagorean theorem to ΔEGI, giving us $EI^2 = GI^2 + GE^2$. Put another way: $(2 - r)^2 = r^2 + (\sqrt{5} - 1)^2$; therefore, $4 - 4r + r^2 = r^2 + 5 - 2\sqrt{5} + 1$. This determines the length of the radius of the inscribed circle $r = \frac{\sqrt{5}-1}{2} = \frac{1}{\phi}$. Now with some simple substitution we have $EM = 2$ and $HM = 2r$, yielding the ratio $\frac{EM}{HM} = \frac{2}{2r} = \frac{1}{r} = \phi$.

Golden Ratio Construction 8

A more contrived construction also yields the Golden Section of a line segment. We will construct a unit square with one vertex placed at the center of a circle whose radius is the length of the diagonal of the square. On one side of the square, we will construct an equilateral triangle, as we show in Figure 3-11. By applying the Pythagorean theorem to triangle ACD, we get the radius of the circle as $\sqrt{2}$, which gives us the lengths of AD, AG, and AJ. By symmetry, we get $BH = CF = x$. Again, applying the theorem involving intersecting chords of a circle (as in Construction 6), we get the following: $GB \cdot BJ = HB \cdot BF$, which gives us $(\sqrt{2}+1)(\sqrt{2}-1) = x(x+1)$. Therefore, $x = \frac{\sqrt{5}-1}{2}$. Once again, we find the segment BF is partitioned

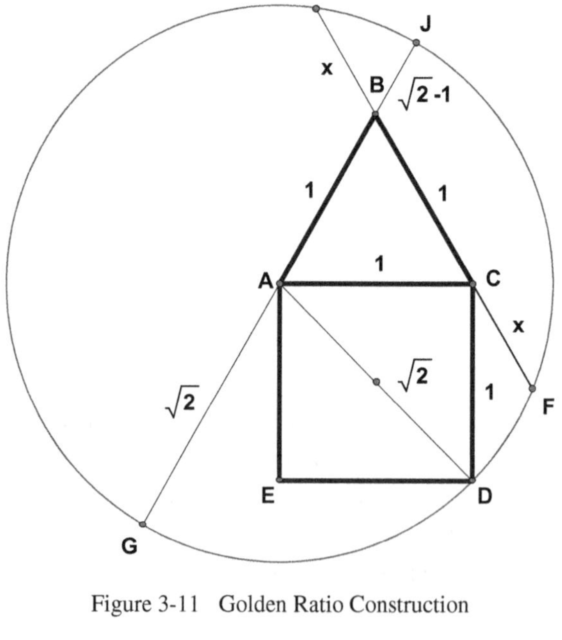

Figure 3-11 Golden Ratio Construction

into the Golden Section at point C, since $\frac{BC}{CF} = \frac{1}{x} = \frac{2}{\sqrt{5}-1} = \frac{\sqrt{5}+1}{2} \approx$ 1.61803398874989484820458683436564, which we recognize as the value of the Golden Ratio.

Golden Ratio Construction 9

We can derive the equation $x^2 + x - 1 = 0$, the so-called *Golden Equation*, in a number of other ways. We construct a circle with a chord AB, which is extended to a point P so that when a tangent from P is drawn to the circle, its length equals that of AB. We can see this in Figure 3-12, where $PT = AB = 1$. Here we will apply a geometric theorem that states that when, from an external point P, a tangent (PT) and a secant (PB) are drawn to a circle, the tangent segment is the mean proportional between the entire secant and the external segment, that is, $\frac{PB}{PT} = \frac{PT}{PA}$. This yields $PT^2 = PB \cdot PA$, or $PT^2 = (PA + AB) \cdot (PA)$. If we let $PA = x$, then $1^2 = (x + 1)x$, or $x^2 + x - 1 = 0$, and as before we can conclude that point A determines the Golden Section of line segment PB, since the solution to this equation is the Golden Ratio.

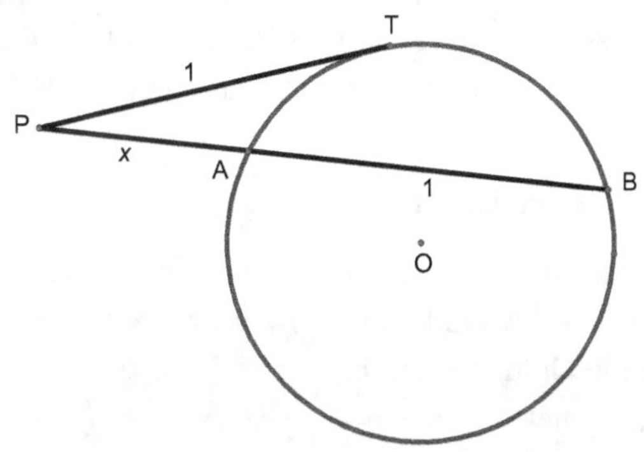

Figure 3-12 Golden Ratio Construction

Golden Ratio Construction 10

This time we will construct the Golden Section with three equal circles. Consider the three tangent congruent circles with radius $r = 1$, as shown in Figure 3-13. Here, we have $AE = 2$ and $BE = 4$. We apply the Pythagorean theorem to ΔABE to get $AB = \sqrt{2^2 + 4^2} = \sqrt{20} = 2\sqrt{5}$. Because of the

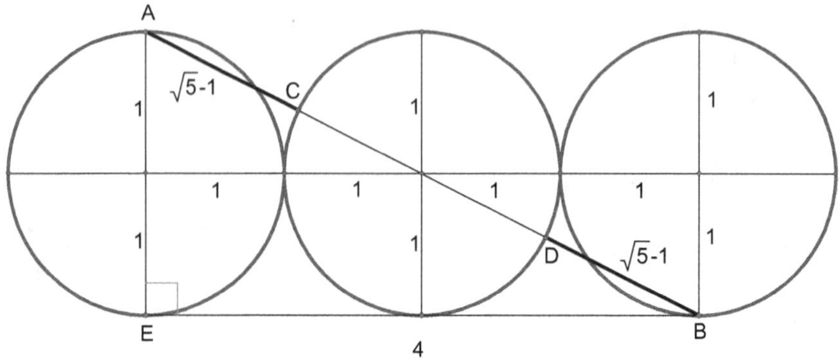

Figure 3-13 Golden Ratio Construction

symmetry in $AC = BD$ and $CD = 2$, we then have $AB = AC + CD + BD = 2AC + BD = 2AC + 2$. Therefore, $2AC + 2 = 2\sqrt{5}$. It then follows that $AC = \sqrt{5} - 1$ and $AD = AB - BD = AB - AC = 2\sqrt{5} - (\sqrt{5} - 1) = \sqrt{5} + 1$. The ratio $\frac{AD}{CD} = \frac{\sqrt{5}+1}{2} \approx 1.61803398874989485$ again denotes the Golden Ratio.

You may have noticed that we have been using a unit measure as our basis. We could have used a variable, such as x, and we would have gotten the same result; however, using 1 rather than x is just a bit simpler.

Golden Ratio Construction 11

When we place the three equal unit circles tangent to each other and tangent to the semicircle as shown in Figure 3-14, we have the makings for another construction of the Golden Section.

First, we note that $AM = BM = JM = KM = LM = R$; and $GH = GM = CE = DF = r = 1$; $CM = DM = EG = FG = 2$; and $EM = R - r = R - 1$. When we apply the Pythagorean theorem to ΔCEM in Figure 3-14, we get $(EM)^2 = (CM)^2 + (CE)^2$, or $(R - 1)^2 = 2^2 + 1^2$. When we solve this equation for R, we get $R^2 - 2R + 1 = 5$, and then $R = 1 \pm \sqrt{5}$. Since a radius cannot be negative, we only use the positive root of R; therefore, $R = 1 + \sqrt{5}$. We then take the ratio $\frac{R}{r} = \sqrt{5} + 1$. Half of this ratio will give us the Golden Ratio: $\frac{1}{2}\left(\frac{R}{r}\right) = \frac{\sqrt{5}+1}{2}$. Therefore, $\frac{LM}{HM} = \frac{R}{2r} = \frac{R}{2} = \frac{\sqrt{5}+1}{2} \approx 1.61803398874989485$. Additionally, the ratios $\frac{HM}{HJ}$ and $\frac{CM}{AC}$ also produce the Golden Ratio,

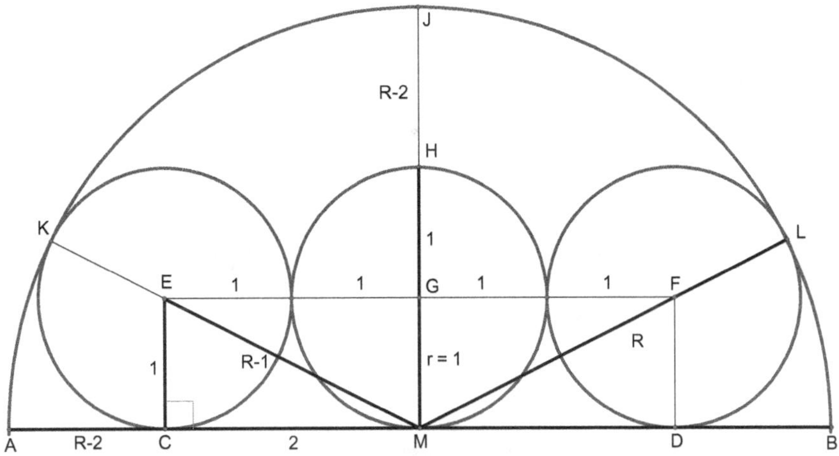

Figure 3-14 Golden Ratio Construction

since with $R - 2r = R - 2 = 1 + \sqrt{5} - 2 = \sqrt{5} - 1$, which then gives us $\frac{HM}{HJ} = \frac{CM}{AC} = \frac{2r}{R-2r} = \frac{2}{\sqrt{5}-1} = \frac{\sqrt{5}+1}{2}$.

Golden Ratio Construction 12

The last in our collection of constructions of the Golden Section is one that may look a bit overwhelming but actually is very simple, as it uses only a compass! All we need is to draw five circles.[4] In Figure 3-15, we begin by constructing circle c_1 with center M_1 and radius $r_1 = r$. Then, with a randomly selected point M_2 on circle c_1, we construct circle c_2 with center M_2 and radius $r_2 = r$; naturally, $M_1M_2 = r$. We indicate the points of intersection of the two circles, c_1 and c_2, as A and B. Constructing circle c_3 with center B and radius $AB = r_3$ will intersect circles c_1 and c_2 at points D and C, respectively. We now construct circle c_4 with center at M_1 and radius $M_1C = r_4 = 2r$. Lastly, circle c_5 with center M_2 and radius $M_2D = r_5 = r_4 = 2r$ is constructed so that it intersects circle c_4 at points E and F.

[4]Kurt Hofstetter, "A Simple Construction of the Golden Section," *Forum Geometricorum* 2 (2002), 65–66.

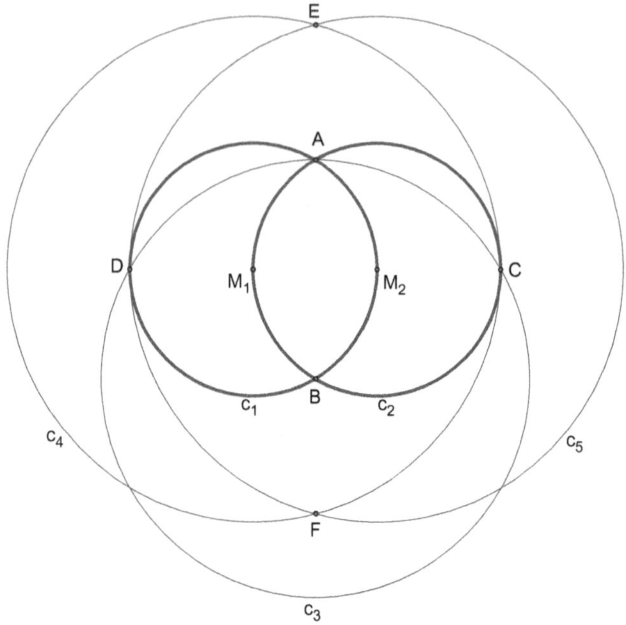

Figure 3-15 Golden Ratio Construction

From Figure 3-16, as a result of obvious symmetry, $AE = BF$, $AF = BE, AM = BM, EM = FM, CM = DM$, and $MM_1 = MM_2$. We can then get $\frac{AB}{AE} = \frac{BE}{AB} = \phi$ (or analogously, $\frac{AB}{BF} = \frac{AF}{AB} = \phi$). This can be justified by inserting a few line segments. The radius of the first circle is $r_1 = r = AM_1$, and the radius of the fourth circle is $r_4 = 2r = CM_1 = EM_1$. We can apply the Pythagorean theorem to ΔAMM_1 to get $AM_1^2 = AM^2 + MM_1^2$, or $r^2 = AM^2 + (\frac{r}{2})^2$, which then determines $AM = \frac{r}{2}\sqrt{3}$. By applying the Pythagorean theorem to ΔEMM_1, we get $EM^2 + MM_1^2 = EM_1^2 = CM_1^2$, or $(2r)^2 = EM^2 + (\frac{r}{2})^2$, whereupon $EM = \frac{r}{2}\sqrt{15}$.

We now need to show that the ratio we asserted above is, in fact, the Golden Ratio.

$$\frac{AB}{AE} = \frac{AM + BM}{EM - AM} = \frac{2AM}{EM - AM} = \frac{2 \cdot \frac{r}{2}\sqrt{3}}{\frac{r}{2}\sqrt{15} - \frac{r}{2}\sqrt{3}} = \frac{2\sqrt{3}}{\sqrt{3}(\sqrt{5} - 1)}$$

$$= \frac{2}{\sqrt{5} - 1} \cdot \frac{\sqrt{5} + 1}{\sqrt{5} + 1} = \frac{\sqrt{5} + 1}{2} = \phi.$$

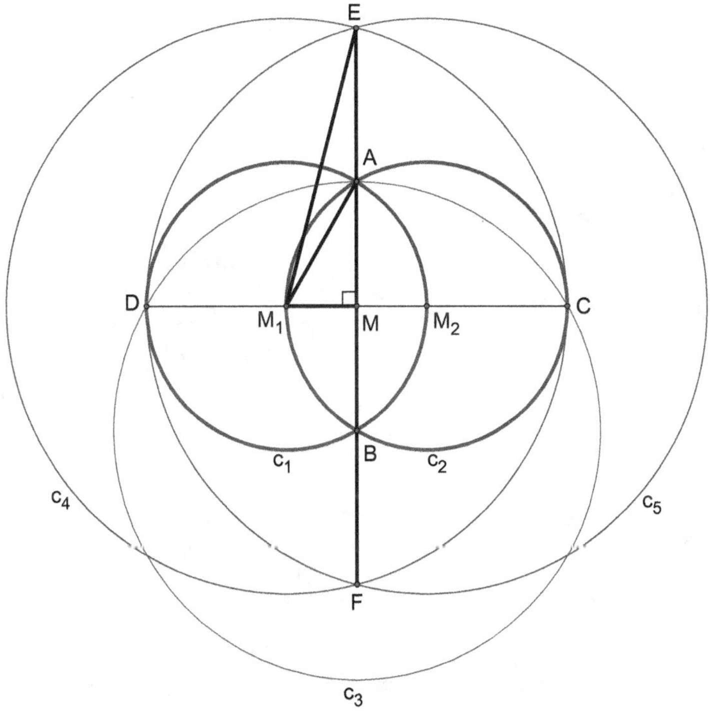

Figure 3-16 Golden Ratio Construction

Now the second ratio that we must check is

$$\frac{BE}{AB} = \frac{EM + BM}{AM + BM} = \frac{EM + AM}{2AM} = \frac{\frac{r}{2}\sqrt{15} + \frac{r}{2}\sqrt{3}}{2 \cdot \frac{r}{2}\sqrt{3}} = \frac{\sqrt{3}(\sqrt{5}+1)}{2\sqrt{3}}$$

$$= \frac{\sqrt{5}+1}{2} = \phi.$$

In both cases we have shown that the Golden Ratio is in fact determined by the five circles we constructed.

Chapter 4

The Aesthetics of the Golden Rectangle in Architecture

The search for the perfect proportion has occupied mankind for centuries, but this problem will probably never be solved to everyone's satisfaction. After all, the side lengths of a (rectangular) object and its proportions are not only subject to visual subjective perception, but are also determined by other external influences, such as purpose, material, the availability of space, etc. Consider a door, for example: it is more important for the door to allow passage through it than to be perceived as aesthetic in its shape.

In other disciplines, however, such factors do not play a significant role. Here one can decide for oneself what shape the desired object should have. So, the external impression is crucial. What size, what proportions does an object need to have so that it looks as beautiful and balanced as possible? Artists and architects have always been concerned with this question. After all, aesthetic perception is a purely subjective matter that differs from person to person and, therefore, cannot be generalized or even made measurable. Or can it be?

Let's start with a thought experiment: we must choose a poster, a photo, a canvas, or some similar wall decoration. First of all, we should consider which format, that is, which aspect ratio the image should have. We decide to use a rectangular shape, since a square picture on the wall might look unbalanced or "bulky." Rectangles usually appear more balanced and "calmer" to the observer's eye than squares with their four sides of equal length. The more the aspect ratio of the rectangle approaches the square's ratio of 1:1, the less satisfying the rectangle appears to the eye. However, the longer and narrower the rectangle becomes, the less appealing it is to the viewer. The optically ideal rectangle seems to lie somewhere in between.

In 1876, the German philosopher Gustav Theodor Fechner (1801–1887) surveyed a large population to determine the dimensions of the most visually appealing rectangle. Although this research was conducted a long time ago, nothing has changed in terms of its significance and relevance. Fechner presented his subjects with a series of ten rectangles with different aspect ratios, from 1:1 (square) to 2:5, including one close to the Golden Rectangle. They were then asked to indicate which one they thought was the most appealing. There were no specifications about which criteria they should use to make their decisions or what the rectangle should be used for. They had to answer intuitively which rectangle impressed them the most with its beauty. Figure 4-1 shows the ten rectangles and their aspect ratios. The one with ratio **21:34** is closest to the Golden Rectangle, and received the most votes.

The further a rectangle deviated from the Golden Rectangle in the direction of a square or in the direction of more extreme aspect ratios, the less

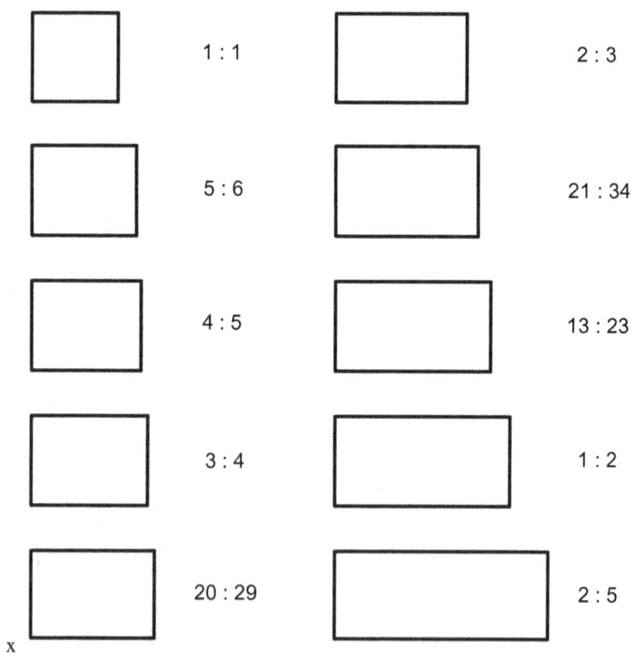

Figure 4-1 Rectangles by Fechner

popular it was. So, it was shown that there is indeed a best-proportioned rectangle that intuitively appeals to the majority of people. For this reason, Golden Rectangles and the Golden Ratio can also be found in a large number of old artworks, including paintings, sculptures, and architecture. The *Mona Lisa* by Leonardo da Vinci (Paris, France), a self-portrait of Albrecht Dürer (Munich, Germany), and the *Sistine Madonna* by Raphael (Dresden, Germany), shown in Figure 4-2, all exhibit the Golden Ratio.

Figure 4-2 Golden Ratio in Famous Artworks

Historic Examples

Great Pyramid of Giza

The Great Pyramid of Giza near Cairo, Egypt, shown in Figure 4-3, is the only one of the Seven Wonders of the Ancient World that still exists today. It is one of the most famous and most visited buildings in the world, probably because many myths and legends are entwined around its construction, purposes, and existence. The pyramid's enormous size and geometrical precision have even spawned conspiracy theories that it could not have been built by human hands, but must have been the work of extraterrestrial civilizations.

The Great Pyramid, also known as the Cheops-Pyramid, is the largest of the three pyramids at Giza and was constructed according to strict astronomical and mathematical specifications. Its position, and its astonishingly precise orientation according to the cardinal points, dimensions, angles, etc., were chosen in such a way that there are even theories of the pyramids *mirroring* star-constellations in the night sky. However, these properties will not

Figure 4-3 Great Pyramid of Giza
Source: Nina Aldin Thune, CC-BY-SA 3.0.

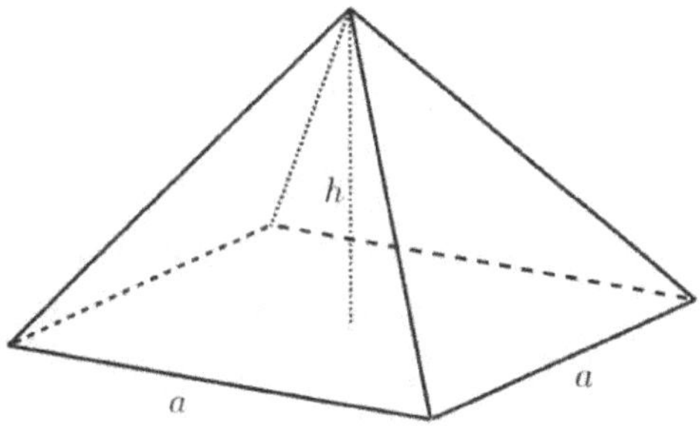

Figure 4-4 $a = 232.81\,m$, $h = 148.21\,m$ Geometric Depiction of the Pyramid

be discussed further here, since they would deviate too far from the actual topic of this section. We shall instead consider the the Pyramid of Giza's geometric properties. Start by noting the measurements and dimensions of the pyramid shown in Figure 4-4.

Even with only these two measurements, we can see an interesting relationship. To do this, we divide the height of the pyramid by half the side length of the base square to get

$$\frac{h}{\frac{a}{2}} = \frac{148.21}{116.405} \approx 1.2732.$$

This number squared gives $1.2732^2 \approx 1.621$, a value that is surprisingly close to ϕ. If we cut the pyramid in half, perpendicular to a base edge, it will result in a triangle ABC, as shown in Figure 4-5. It gets divided in half into two (almost) congruent right triangles AMC and BMC by its height h.

Knowing the length of h and $\frac{a}{2}$, we can easily calculate the length of the hypotenuse c (height of a lateral triangle): $c = \sqrt{h^2 + \left(\frac{a}{2}\right)^2} = \sqrt{148.21^2 + 116.405^2} \approx 188.46\,m$. This number leads us to another interesting relation between the measurements. Divide the hypotenuse, c, by the pyramid's height, h:

$$\frac{c}{h} = \frac{188.46}{148.21} \approx 1.2716.$$

Curiously, this value is also very close to $\sqrt{\phi}$. We see that the proportion between the height of the pyramid and the base lengths is nearly the same as the proportion between the height of a lateral triangle and the height of the

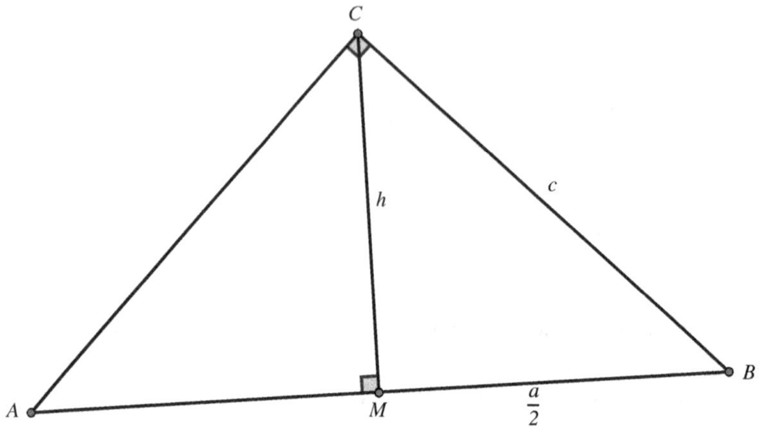

Figure 4-5 Half the Base Area of the Pyramid

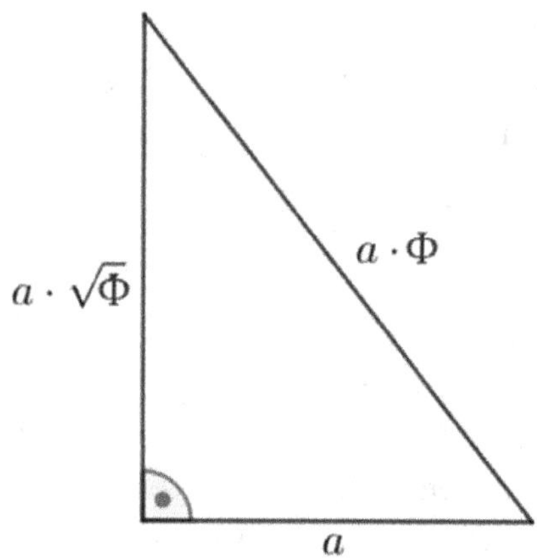

Figure 4-6 A Special Cut Through the Pyramid

pyramid. And it shows a strong relationship to the Golden Ratio, as shown in Figure 4-6. Furthermore, consider the ratio between the hypotenuse c (height of a lateral triangle) and half its base length $\frac{a}{2}$ of the pyramid: $\frac{c}{\frac{a}{2}} = \frac{188.46}{116.405} \approx 1.619$, which is also close to ϕ.

Only right triangles with the measurements shown in Figure 4-6 ($a \in \mathbb{R}^+$) have the property of the quotient of the hypotenuse and the longer leg being the same as the quotient of the longer leg and the shorter leg. The Great Pyramid of Giza is constructed with such a triangle within its cut-surface. Here we offer an explanation. We are looking for a right triangle with the legs x (shorter) and y (longer) and hypotenuse z, where $\frac{z}{y} = \frac{y}{x}$. By incorporating the Pythagorean theorem, we get the following equation system:

$$z = \sqrt{x^2 + y^2}, \tag{1}$$

$$\frac{z}{y} = \frac{y}{x}. \tag{2}$$

Substituting z into equation (2), we get $\frac{\sqrt{x^2+y^2}}{y} = \frac{y}{x}$. We rewrite the equation's left side step-by-step: $\frac{\sqrt{x^2+y^2}}{y} = \sqrt{\frac{x^2+y^2}{y^2}} = \sqrt{\frac{x^2}{y^2} + 1}$ $\Longrightarrow \sqrt{(\frac{x}{y})^2 + 1} = \frac{y}{x}$. We square both sides: $(\frac{x}{y})^2 + 1 = (\frac{y}{x})^2$. Now we will substitute λ for the term $(\frac{y}{x})^2$ into the previous equation to get $\frac{1}{\lambda} + 1 = \lambda$. We know from Chapter 3 that the only positive solution for λ in this equation is $\lambda = \phi$. So $(\frac{y}{x})^2 = \phi$, which leads us back to $\frac{z}{y} = \frac{y}{x} = \sqrt{\phi}$, so that $y = x\sqrt{\phi}$ and $z = x \cdot \phi$. Besides that, there is another way to explain these numbers in the pyramid's measurements, regardless of ϕ.

The two values mentioned above are not only close to $\sqrt{\phi}$, but also close to $\frac{4}{\pi}$ (≈ 1.2732). From today's point of view, it is difficult to determine which of these two relations was employed for the construction of the pyramid, as the exact original measurements of the pyramid in ancient days are lost. Another indication of the use of π, however, lies in the surface areas: the base area is $a^2 = 232.81^2 \approx 54200.5\,m^2$, and the cut-surface area in Figure 4-5 is $\frac{a \cdot h}{2} = \frac{232.81 \cdot 148.21}{2} \approx 17252.39\,m^2$. So, the quotient is $\frac{54200.5}{17252.39} \approx 3.14162$. This number corresponds to π in its three first decimal-places.

Both theories, the one with ϕ and the one with π, have their advocates. It just seems like a curious coincidence, that $\sqrt{\phi}$ and $\frac{4}{\pi}$ are so similar to each other.

Parthenon at Athens

One of the famous examples of ancient architecture is the *Parthenon* on top of the Acropolis in the heart of Greece's capital city, Athens, shown in Figure 4-7. Not only does it exhibit prototypical elements of traditional ancient Mediterranean sacral architecture, such as tall columns shaping a

Figure 4-7 Acropolis in Athens, Greece
Source: Kostas Vourou, Unsplash.

rectangle with a triangular gable at the top (*tympanum*), but the temple's measurements hide a lot of interesting geometry and mathematics.

Archeologists have not reached a consensus on whether the geometric properties behind the structure were purposely calculated before construction, or just generated by chance through a mixture of luck and good taste. However, we will still have a glimpse at these mathematical peculiarities. The floor of this temple is 228 ft long and 101 ft wide. So, the ratio of this rectangle is $\frac{228}{101} \approx 2.257$. We will need this number a bit later. First, we will consider the construction shown in Figure 4-8, based on two Golden Rectangles. Construct a Golden Rectangle (red) and then add another Golden Rectangle (blue) adjacent to it, whose longer side matches the larger rectangle's shorter side.

The whole combined rectangle has a ratio of ≈ 2.236, which is almost identical to the ratio of the floor of the Parthenon. They differ by less than 1%. We offer the following explanation. The exact measurements of the red Golden Rectangle do not matter, so we let the shorter side have length

Figure 4-8 Two Golden Rectangles

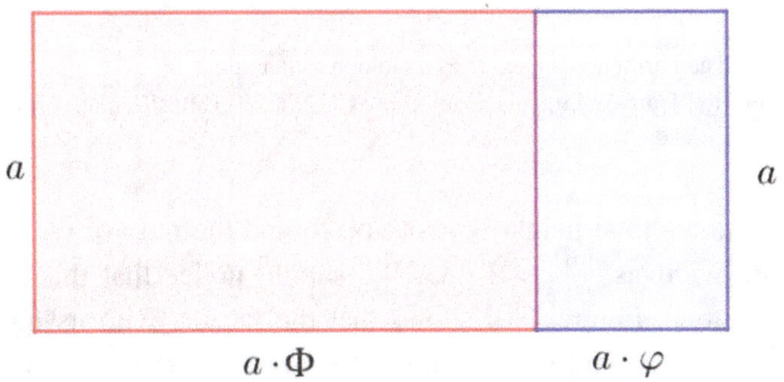

Figure 4-9 Measurements of the Golden Rectangles

$a(a \in \mathbb{R}^+)$; therefore, its longer side is $a \cdot \phi$. Viewing the blue rectangle, we see that its longer side has length a, so its shorter side calculates as $a \cdot \varphi$. Figure 4-9 shows the measurements of the Golden Rectangles.

In total, this construction has edges with lengths $(a \cdot \phi + a \cdot \varphi)$ and a. The ratio of the sides is

$$\frac{a \cdot \phi + a \cdot \varphi}{a} = \frac{a \cdot (\phi + \varphi)}{a} = \phi + \varphi \approx 1.618 + 0.618 = 2.236.$$

The height between the Parthenon's upper step and the lower edge of the gable is about 45 ft. So, when looking at the front of the building without the gable, it fits into a rectangle whose sides are in the ratio $\frac{101}{45} = 2.2\dot{4}$. This value is very near the other two ratios mentioned earlier. So, the (almost) rectangular part of the façade shows a ratio similar to that of the floor.

Figure 4-10 The Parthenon's Façade as a Golden Rectangle
Source: Modified from Wikimedia Commons, CC BY 3.0 (https://creativecommons.org/licenses/by/3.0/deed.en).

The Parthenon's total height is about 60 ft, and the ratio of the building's height and width is $\frac{101}{60} \approx 1.68\dot{3}$. We should notice that this number is quite close to ϕ. Figure 4-10 shows that the façade, with its figuratively reconstructed pediment (dashed line), fits almost perfectly into a Golden Rectangle (bold outline).

The outer rectangle of the front-façade including the steps measures 111.31 ft wide and 65.162 ft high. So, it has a ratio of $\frac{111.31}{65.162} \approx 1.7082$. A rectangle with this ratio can be exactly divided into the following: A square and three additional rectangles with the ratio $\sqrt{5} - 2$, which is the same as $2 \cdot \varphi - 1$, as shown in Figure 4-11. The Golden Ratio occurs multiple times in the Parthenon's façade. This can be explained as follows: The rectangle shown in Figure 4-11 has measurements of $a (a \in \mathbb{R}^+)$ and $a + 3a(2\varphi - 1) = a(6\varphi - 2)$. Its ratio can be calculated as $\frac{a \cdot (6\varphi - 2)}{a} = 6\varphi - 2 \approx 1.7082$.

Let A and B be the endpoints of the horizontal baseline of the ground floor. The points C and D are perpendicularly above the points A and B at the lower edge of the gable, thereby creating the rectangle $ABCD$, as shown in Figure 4-12. This rectangle consists of exactly four Golden Rectangles placed side-by-side, so that rectangle $ABCD$ has a side-ratio of $4 \cdot \varphi$.

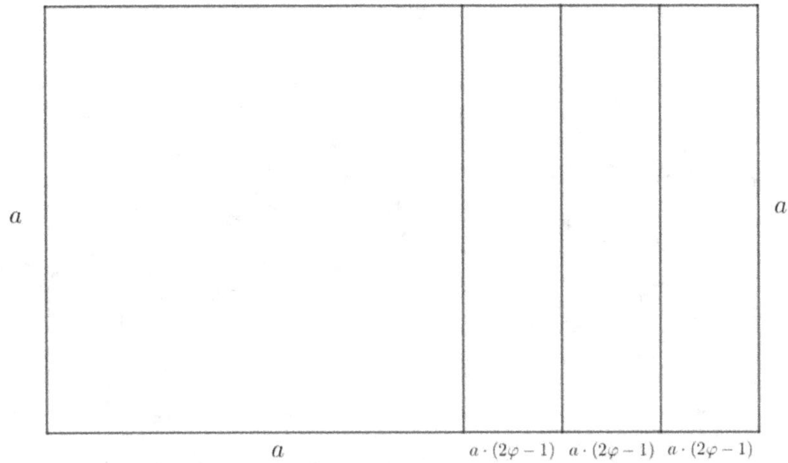

Figure 4-11 The Number φ in the Parthenon's Façade

Figure 4-12 Golden Rectangles in the Parthenon's Façade (1)

Now scale down *ABCD* to a similar rectangle *EFGH*, shown in Figure 4-13, where *EF* is still the horizontal baseline of the ground floor and *GH* is defined by the upper edge of the columns. Due to its similarity to *ABCD* (side-ratio $= 4 \cdot \varphi$), it can also be subdivided into four Golden Rectangles placed side-by-side. This newly generated rectangle helps us understand the position of the columns. They were not placed randomly, but follow strict geometrical requirements related to φ.

Figure 4-13 Golden Rectangles in the Parthenon's Façade (2)

Figure 4-14 Golden Rectangles in the Parthenon's Façade (3)

Let's consider one of the four Golden Rectangles and divide it into 3×3 congruent smaller rectangles—which are, of course, all Golden as well. We can subdivide *EFGH* into 12×3 congruent Golden Rectangles. Figure 4-14 shows the 11 vertical separation lines. It is now discernible that the inner columns are always spaced exactly two lines apart.

Suppose we let the height of a column $EH = h$. Then $EF = 4\varphi h$. The distance between two columns is one-sixth of this length, that is, $\frac{2}{3}\varphi h$. That means that the distance between two columns is $\frac{2}{3} \cdot \varphi \approx 0.412$ times

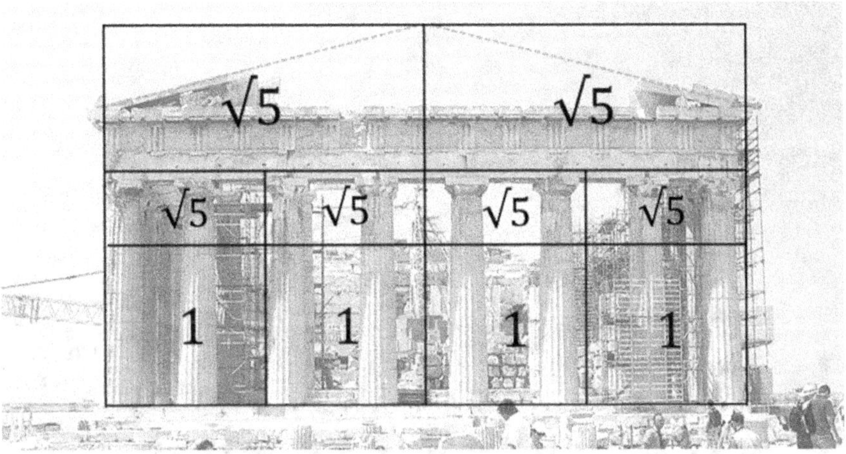

Figure 4-15 Subdivision of the Pantheon's Façade

their height. But the front façade of the Parthenon can be further subdivided, revealing geometrical peculiarities related not to φ, but to $\sqrt{5}$. The omnipresence of $\sqrt{5}$ is illustrated by Figure 4-15, where the numbers inside the rectangles indicate their ratio. The bottom line of the *epistyle* (the long beam lying on top of the columns) divides the rectangle into an upper part, consisting of the epistyle and the gable, and a lower part, consisting of the columns. The whole upper part is composed of two identical rectangles with ratio $\sqrt{5}$. Only the small outmost sides of the gable on the left and on the right side are outside of these rectangles. The lower part of the rectangle can be divided vertically into four smaller congruent rectangles, each of which has the proportion of $\frac{1}{\sqrt{5}} + 1$, so every single one of these four rectangles can be cut into a square and a rectangle with ratio $\sqrt{5}$ on top of it.

Taj Mahal

The *Taj Mahal*, shown in Figure 4-16, is an impressive monument on the banks of the Yamuna River in Agra, Uttar Pradesh, in northern India. It was built between 1632 and 1648 as a mausoleum for Shah Jahan's wife Mumtaz Maha. The Shah wanted to provide a magnificent final resting place for his wife, who died in 1631, but he also intended to show his power and

Figure 4-16 The Taj Mahal

the prosperity of his rule. Both purposes were fulfilled, as Jahan's historian Muhammad Amin Qazwini stated during the construction-process:

> *And a dome of high foundation and a building of great magnificence was founded—a similar and equal to it the eye of the age has not seen ..., and of anything resembling it the ear of time has not heard in any of the past ages ... it will be the masterpiece of the days to come, and that which adds to the astonishment of humanity at large.*

To this day, the Taj Mahal counts as the most famous and greatest building in Indo-Islamic culture and charms visitors from all over the world. To preserve this architectural milestone for future generations, the Taj Mahal was designated as a World Heritage site by UNESCO in 1983.

Consider how geometrically elaborate the whole complex is. The Shah designed the building using grids of a length unit called *gaz* throughout, with the modules of the complex strictly arranged in this grid. Symmetry was crucial. Nothing was left to chance. There is much information available about the architectural layout, symbols, and components of the Taj Mahal and how all these criteria lead to its designation as a World Heritage site. However, many sources do not acknowledge the presence of the Golden

Ratio anywhere in the whole complex. In contrast to the other buildings mentioned earlier, there seems to be no mathematical evidence of the purposeful integration of the Golden Ratio into the Taj Mahal's dimensions. But is this really true?

The observer can still find "hidden," yet conspicuous, Golden Rectangles. From an observation point in front of the façade, right at the axis of symmetry, where one has the best view of the Taj Mahal and where most of the popular pictures of the building are taken, one can see that the base wall, the front minarets, and the tip of the central dome outline a Golden Rectangle that frames the building, as shown in Figure 4-17. Furthermore, when partitioning a square from this Golden Rectangle (which, as we now know, generates another Golden Rectangle), the side of this square meets the outer edge of the frontal entrance component, as can be seen in Figure 4-17, shown by the thick dotted line.

We repeat this step in the new Golden Rectangle and partition a square from its bottom portion to provide an additional line, which nearly meets the top of the smaller domes next to the big central dome, as shown by the thin dotted line in Figure 4-17. A closer look at the central main portal reveals even more interesting geometrical structures, which we can see in Figure 4-18.

The upper edge of the base wall, the vertical side-edges of the portal, and the lower edge of the ornamental decorative calligraphy enclose a Golden Rectangle (thick black outline). So, when looking at the Taj Mahal, one's view is directed to a large Golden Rectangle in the middle of the building. Dividing this Golden Rectangle into a square and leaving another Golden Rectangle generates a horizontal line, which meets the lower edge of the horizontal bar right beneath the central window. Demarcating another square provides a vertical line that corresponds to the outer edge of the central door. So, the width of this door fulfills the requirements of "golden geometry." One last iteration: Cutting off a square at the lower-right corner provides another horizontal line. Extended over the whole width of the portal, this dotted line nearly touches the upper tip of the central door. Not only does the door's width correspond to golden geometry, but also its height is related to Golden Rectangles. These were the geometrical structures in the portal when starting by separating a square from the upper part of the Golden

Figure 4-17 Golden Rectangles in the Taj Mahal's Façade

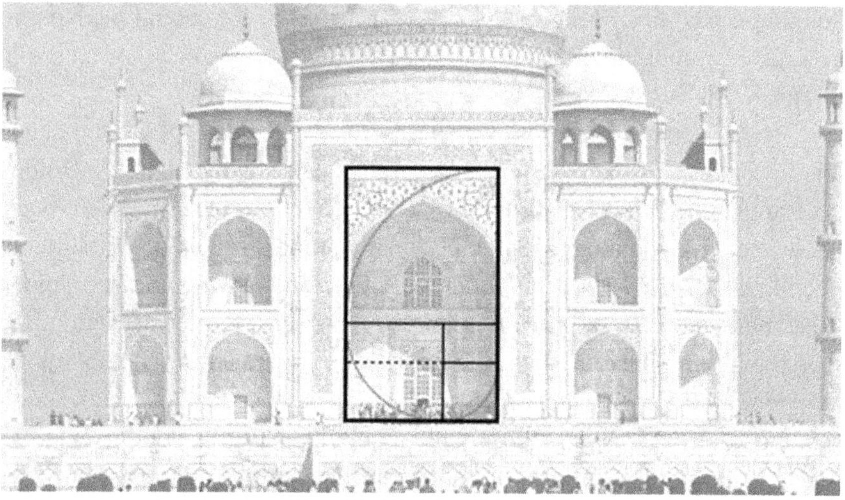

Figure 4-18 Golden Rectangles in the Taj Mahal's Portal (1)

Rectangle, as we have seen in Figure 4-18. But can anything be discovered when the initial square is separated from the bottom? Using Figure 4-19, we should find the answer.

Figure 4-19 Golden Rectangles in the Taj Mahal's Portal (2)

The newly generated horizontal line (thin line) is precisely in contact with the upper tip of the central window. Thus, all in all, the most prominent and popular view of the Taj Mahal offers a large variety of golden geometrical structures. Not only is the whole complex of the central mausoleum and the minarets enclosed by a Golden Rectangle, but also the structure's portal is geometrically peculiar. The central door, the central window, and the entire structure relate through golden geometry. Maybe the exact measurements of the components mentioned above are not exactly in the Golden Ratio. But thanks to the perspective distortion of how the observers see them from the viewpoint, to the eye they appear to be in Golden relations.

Other Examples from Around the World

Temple of Kukulcán

The Temple of Kukulcán, shown in Figure 4-20 and better known as *El Castillo*, is a famous pyramid located in the ancient Maya city of Chichén Itzá in the north of Mexico's Yucatán Peninsula. Built between the 11th and 13th centuries, *El Castillo* is a good example of the knowledge of the Golden Ratio in the pre-Columbian culture of the Maya, since its interior

Figure 4-20 Temple of Kukulcán
Source: David Stanley.

layout displays the Golden Ratio (0.6180339887…), with walls placed so that the outer spaces relate as such to the center chamber.

Notre-Dame de Paris

Although there is no obvious Golden Rectangle to be seen in the west façade of this world-famous gothic cathedral, there is still some interesting hidden golden geometry to be found by standing right in front of the center of the façade. The lower part of the façade (from the floor up to the chimeras) forms a perfect square, as shown in Figure 4-21, where a dotted horizontal line marks the lower edge of the *Galérie des rois* (King's gallery), and another pierces right through the middle of the rose window.

The lower of these lines is at $\frac{3}{8}$ of the square's height, while the second, higher one is at $\frac{5}{8}$ of the square's height. Expanding these fractions to natural numbers, the four horizontal lines of the square mentioned are at height 0,

Figure 4-21 Fibonacci Numbers in the Façade of Notre-Dame, Paris
Source: Daniel Vorndran, CC BY-SA 3.0 (https://creativecommons.org/licenses/by-sa/3.0/).

3, 5, and 8 (Figure 4-21; right-side labels). All these numbers are Fibonacci numbers, and 3, 5, and 8 are consecutive. These prominent horizontal lines have a close relation to the Golden Ratio, since the ratio of consecutive Fibonacci numbers approaches the Golden Ratio as the numbers increase. In addition, the dimensions of the whole façade offer Golden geometry to the observer. The façade is 200 feet high and 128 feet wide, which makes a height-width ratio of 1.5625, somewhat close to the Golden Ratio. We will look at the ratio of the façade from the observer's point-of-view. Thanks to the perspective distortion of the building, a human on the floor perceives the

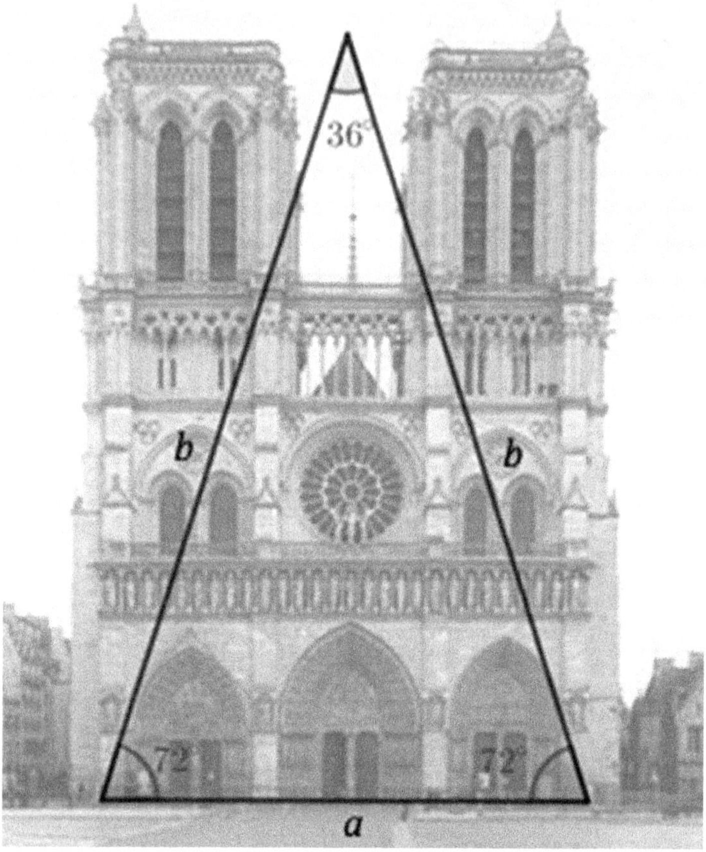

Figure 4-22 Golden Triangle in the Façade of Notre-Dame, Paris

height of the cathedral to be slightly lower, so that, to the eye, the height-side ratio is a bit smaller, about 1.54. This means that if we were to draw an isosceles triangle with the base a being the floor of the church, and the equal sides of length b going up to a central vertex at the upper edge of the towers, the base angles α calculate as $\tan(\alpha) = \frac{1.54a}{0.5a} = 3.08 \Longrightarrow \alpha = \arctan(3.08) \approx 72.01^\circ \approx 72^\circ$, so the triangle has angles of 72°, 72°, and 36°. This is actually a *Golden Triangle*, because $\frac{b}{a} = \phi$, as can be seen in Figure 4-22.

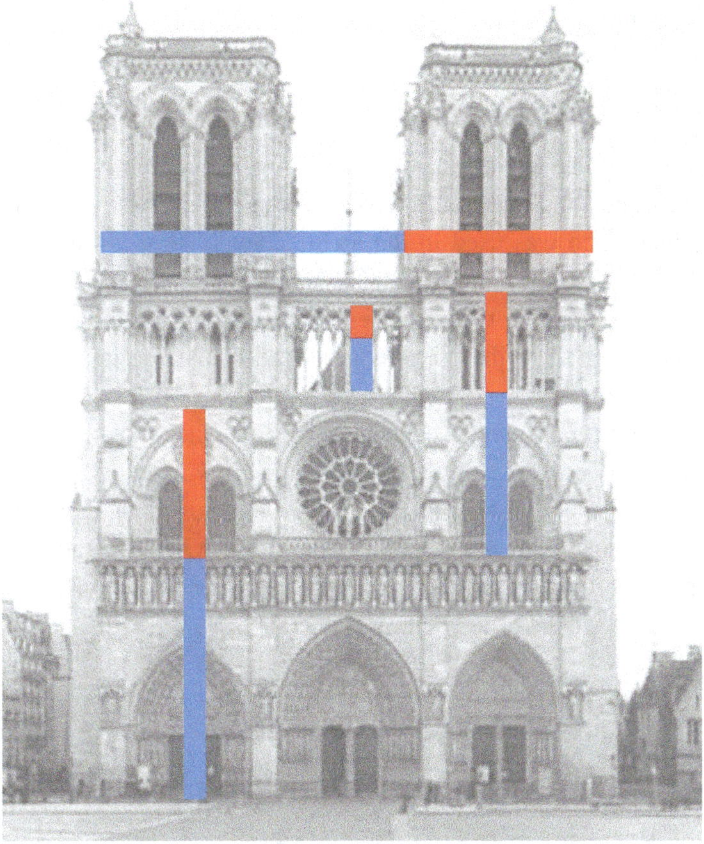

Figure 4-23 Other Golden Ratios in the Façade of Notre-Dame, Paris

Figure 4-23 shows even more examples of the Golden Ratio in the façade. For every blue-red bar, the blue portion is ϕ times longer than the red part, and the whole bar is ϕ times longer than the blue part. So, every bar seen there is divided in the Golden Ratio. It becomes apparent that the whole façade is laid out in such a way that the vertical distances of a multitude of prominent horizontal lines demonstrate the Golden Ratio. A horizontal bar at the top demonstrates that the Golden Ratio can be found at the very top of the church as well. The width of a tower (red part of the bar) corresponds to φ times the width of the cathedral (the whole bar).

St. Peter's Basilica

St. Peter's Basilica in Vatican City is the biggest church in the world and has some interesting interior dimensions. There are some examples of how the Golden Ratio was used in the architecture of this building shown in Figure 4-24. First, the central rectangular part of the façade—between the two columns separating the belltowers from the main building and from the top of the stairs up to the upper edge of the balustrade—forms a perfect Golden Rectangle.

Furthermore, consider the center of the façade, including the main portal. The four central columns topped by the triangular tympanum find their place in a perfect vertical Golden Rectangle that extends upwards to the upper edge of the balustrade, as shown in Figure 4-25. When separating a square from the lower part of the rectangle, the resulting horizontal line exactly meets the upper edge of the columns.

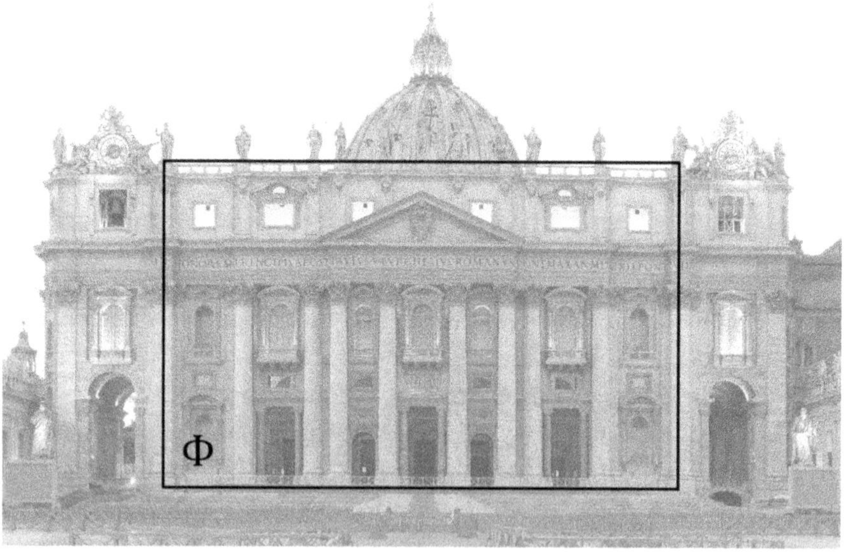

Figure 4-24 Golden Rectangle in St. Peter's Basilica (1)

Figure 4-25 Golden Rectangle in St. Peter's Basilica (2)

Modern and Contemporary Examples

Excursus: Le Corbusier's Modulor

The Swiss architect Charles-Édouard Jeanneret (1887–1965), better known as Le Corbusier, is considered one of the most famous and influential designers of the 20th century. He designed well-known buildings throughout the world. Le Corbusier set milestones in the calculation of ratios and geometry in architecture and design. He developed a system of ratios and measurements called *Le Modulor* with its subtitle *essai sur une mesure harmonique à l'échelle humaine, applicable universellement à l'architecture et à la mécanique* (*A Harmonious Measure for the Human Scale Universally Applicable to Architecture and Mechanics*), where he derived pleasing proportions in design and architecture from statistical measurements of the human body. With this work, he tried to reconcile various other designers and architectural theories. He used a series of measurements from a 1.829 m tall human body with one arm raised. The navel divides the body into two parts with equal lengths, and other prominent joints mark the other dividing

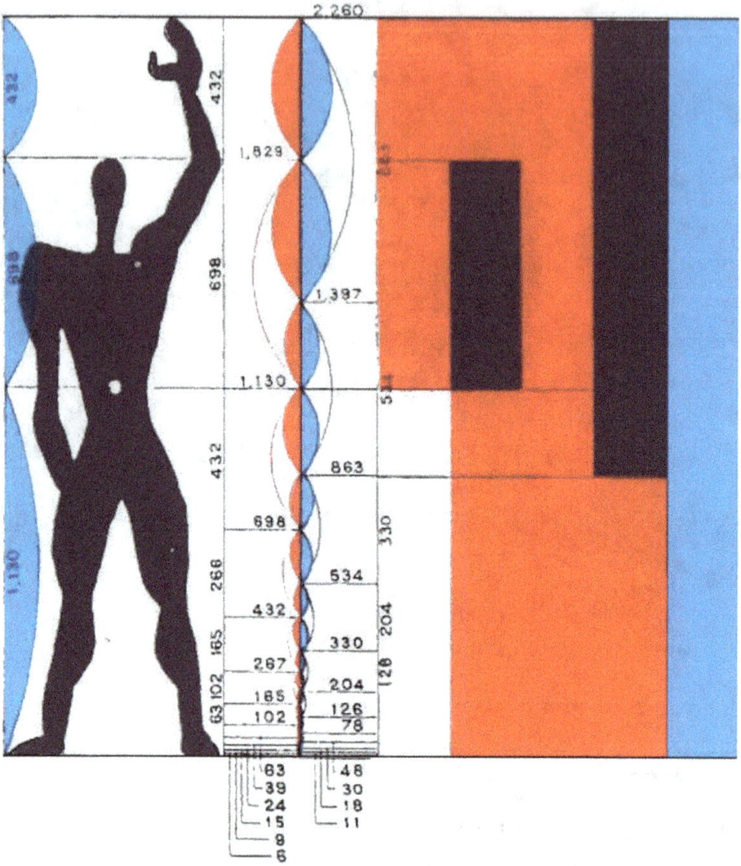

Figure 4-26 Le Modulor

Source: © F.L.C. / *ADAGP*, Paris and *DACS*, London 2024.

points. The exact lengths of the body parts, measured in centimeters, can be seen on the right side of the silhouette in Figure 4-26.

The lengths of the single parts (seen from the bottom upwards) follow a pattern similar to the principle of the Fibonacci sequence: The measurement of one part is always exactly (or at least very close to) the sum of the two parts below it. This means that the ratio of two following body parts is always close to the Golden Ratio. Le Corbusier presented his revolutionary concept to an audience in 1951 in Milan, Italy, where he modestly described it "as if it were an elementary, prosaic tool: 'The Modulor, which I have described to you, is a simple work tool, a tool such as aviation, such as many other improvements created by men.'"[1]

In his architectural work, Le Corbusier made use of the numbers and measurements occurring in *Le Modulor* in various ways. Probably the best-known example for this approach is the residential block *Unité d'Habitation* in Marseille, France, shown in Figure 4-27. In this building, the measurements determined in *Le Modulor* are ubiquitous. The dimensions of the exterior, the interior, and even the furniture—nearly everything is determined to fit the proportions of a human body. The human body served as a natural scale for the building. It seems as though Le Corbusier did not want to leave anything to chance, as he said, "In my works the whole effect should be spread by proportions. I believe in the divinity of the proportion."

Figure 4-27 Unité d'Habitation

Source: CC BY-SA 2.5 https://creativecommons.org/licenses/by-sa/2.5/.

[1] J. L. Cohen, "Le Corbusier's Modulor and the Debate on Proportion in France," *Architectural Histories*, 2:1 (2014).

United Nations Secretariat Building

Since its completion in 1952, the UN Secretariat building has been an integral part of the skyline of New York City. This building was designed and planned by Le Corbusier and the Brazilian architect Oscar Niemeyer (1907–2012). The prominent glass façade of the skyscraper looks like a plain rectangle, but only at first. Delving deeper into the exact dimensions of its components reveals a multitude of hidden applications of the Golden Ratio. It is still disputed whether the conspicuous horizontal separation of the façade was intended to follow the Golden Ratio, as shown in Figure 4-28. All three rectangles (blue, yellow, and red) are Golden Rectangles. The horizontal non-reflective separation bands of the building follow these rectangles in their position and height, but not perfectly. The Golden Rectangles overlap each other. If the architects really intended to design the tower in a perfect Golden Ratio, they wouldn't have had these overlaps.

One possible reason for the deviations from perfect golden proportions is that the Golden Ratio is an irrational number. When calculating the dimensions for such a building, there must be compromises in mathematical precision. Additionally, in contrast to pure mathematical constructions, a building must fulfill certain principles to guarantee its viability. The height of the floors have to follow regulations and provide space for humans. The dimensions of other components must follow static rules depending on their use and the materials used. Thus, the measurements of such a tall building cannot be calculated solely on the basis of desired geometrical properties.

Despite the inevitable inaccuracies of the three Golden Rectangles stacked on each other, there are more applications of the Golden Ratio to be found in the vertical aspects of the façade—very similar to the façade at Notre-Dame in Paris. Just as in Figure 4-23, Figure 4-29 shows two vertical bars, where the blue part is ϕ times longer than the red part and the whole bar is ϕ times longer than the blue part. Thanks to the different number of floors between the horizontal non-reflective bands, some of their distances follow the Golden Ratio. And also the total height of the tower stands in the Golden Ratio to one of these bands. Moreover, there is a green-orange bar near the bottom in Figure 4-29, where the orange part indicates the height

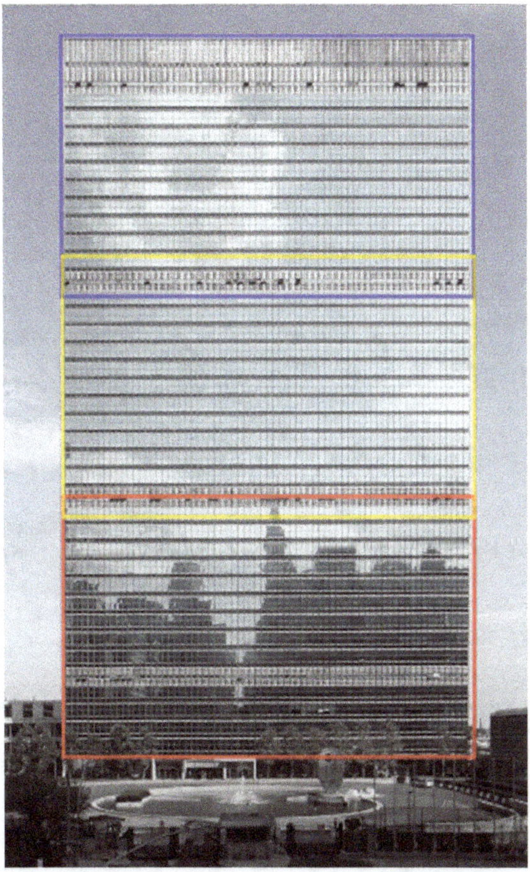

Figure 4-28 Golden Rectangles in the Façade of the UN Secretariat Building

of the entrance, and the green part indicates the height between the entrance and the first horizontal band. These measurements have a relation to ϕ: The green part is ϕ^2 times longer than the orange part.

A more detailed view of the windowpanes of the horizontal separating bands reveals some more Golden Rectangles. The total height of such a band in relation to the width of two window segments shapes a perfect Golden Rectangle, as can be seen in Figure 4-30. Additionally, when cutting off a

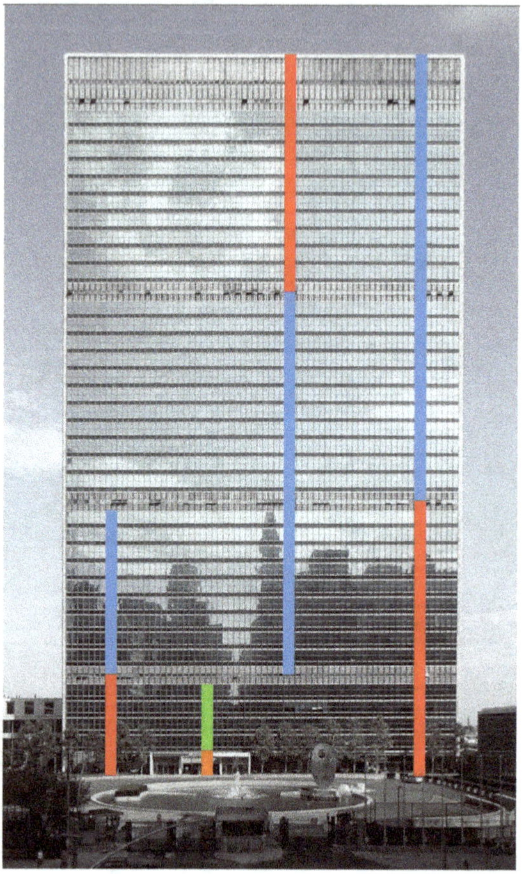

Figure 4-29 Golden Ratios in the Façade of the UN Secretariat Building

square and leaving another Golden Rectangle, the cutting-line exactly meets lower edge of the window-strip, shown by the thin line in Figure 4-30.

Finally, the design of the front entrance of the UN Building offers a visitor lots of Golden geometry. First, the ratio of horizontal distance between the center of the entrance and the first column to the distance between the first and the second column yields the Golden Ratio, as indicated in Figure 4-31 by the same blue-red bar used for Notre-Dame. Second, the lateral large open areas are Golden Rectangles, indicated with the red line in Figure 4-31.

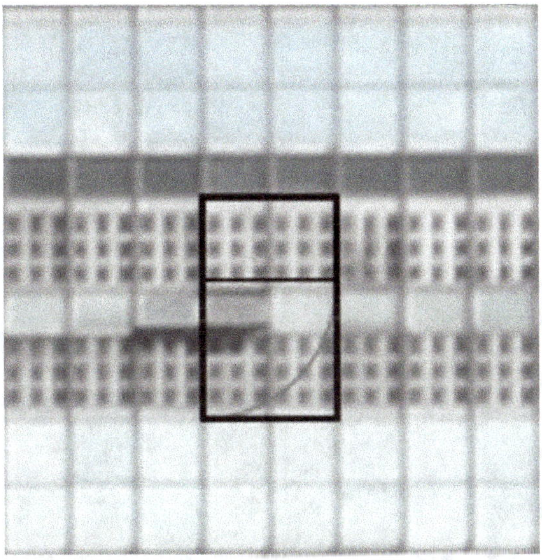

Figure 4-30 Windowpanes as Golden Rectangles

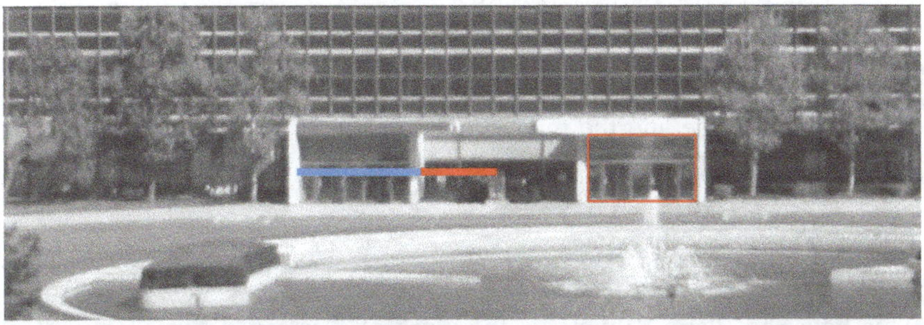

Figure 4-31 Golden Ratios at the Front Entrance

Villa Savoye

We know that Le Corbusier had a particular liking for the Golden Ratio and Golden Rectangles. The Villa Savoye in Poissy, France, is another good example of his typical geometric approach and minimalist architectural style. This building is practical first and foremost, and forgoes any ornamental decoration. Figure 4-32 shows two Golden Rectangles side-by-side, bounded vertically by the ground and roof and horizontally by the outer

Figure 4-32 Golden Rectangles in the Villa Savoye in Poissy, France
Source: Jean-Pierre Dalbéra, Flickr. CC-BY 2.0.

edges of the house and the middle column (distorted due to the camera's perspective).

Farnsworth House

This house in Plano, Illinois, is the work of German American architect Ludwig Mies van der Rohe, who is considered one of the most influential representatives of 20th century modernism. His buildings are characterized by plain shapes, strict geometry, and the use of then-revolutionary and innovative materials like steel and extensive glass panes. Mies van der Rohe was commissioned by Dr. Edith Farnsworth to design and build a weekend retreat, which resulted in this structure, now one of the most prototypical and recognizable works of the architect. Farnsworth, however, was not satisfied with the result and took Mies van der Rohe to court, where she lost her suit. The architect's penchant for geometry made him include Golden Rectangles

Figure 4-33 Golden Rectangles in Farnsworth House
Source: Carol M. Highsmith via Picryl.

in the simple exterior design of the house, which is shown in Figure 4-33. Seen from the outside, the spaces between the columns and between the upper edge of the terrace (dotted line) and the roof form three Golden Rectangles side-by-side (red outlines). Furthermore, the position of the dividing wall in the rectangle on the right is determined by the Golden Ratio. After the third iteration of separating squares from the Golden Rectangle, the dividing line meets the wall, as seen with the red dotted line in Figure 4-33.

Finally, the dimensions of the overhangs to the left (not completely visible in the figure) and right follow the same principle with the yellow Golden Rectangles shown in Figure 4-33. Mies van der Rohe did not only make use of Golden Rectangles in the dimensions of the walls and windows. Figure 4-34 is the architectural floor plan of the house's lower terrace. It reveals that this rectangular shape has a side-ratio of (4φ):1, so that exactly four Golden Rectangles can be placed side-by-side. Furthermore, the position and the width of the two stairs leading from the building onto the terrace, and from the terrace onto the ground, are determined by Golden Rectangles. The western edges of both stairs (on the plan: the left-side edges) follow the separation line of the second iteration of separating square from Golden Rectangle. The eastern edges (on the plan: the right-side edges) follow the separation line of the fourth iteration in their rectangle (red dotted lines).

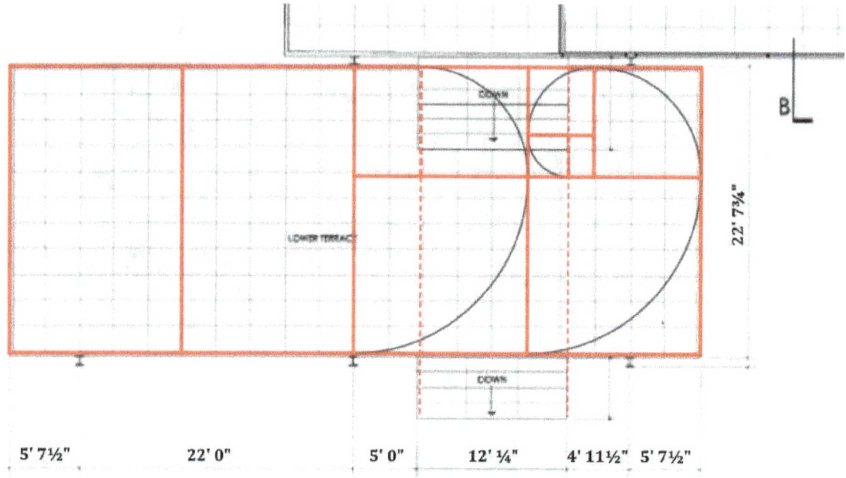

Figure 4-34 Golden Rectangles in Farnsworth House's Lower Terrace

CN Tower

The CN Tower in Toronto, Canada, was the tallest freestanding structure in the world until 2007, when it was overtaken by the Burj Khalifa in Dubai, United Arab Emirates. The total height of the CN Tower is 1,815 ft (553.3 m). Its observation deck is at a height of 1,122 ft (342 m). Dividing the height of the tower by the height of the observation-deck, we get $\frac{1815}{1122} \approx 1.618$: the Golden Ratio ϕ. Figure 4-35 illustrates this ratio, as the blue part of the bar stands in the Golden Ratio to the red part. That the observation deck isn't perfectly aligned with the blue bar is only due to the distortion in perspective distortion from viewing the tower from the ground.

Palacio Barolo

The architect Mario Palanti (1885–1978) included many references to Dante's *Divine Comedy* in his design of the Palacio Barolo in Buenos Aires, Argentina, shown in Figure 4-36. Its tripartite vertical structure alludes to the comedy's three parts: "Hell," "Purgatory," and "Paradise." The number of floors and rooms in each section corresponds to the number of verses in the literary work. And further, probably in reference to the Golden Ratio being called "Divine," there is some golden geometry used in the building's

Figure 4-35 Golden Ratio in the CN Tower

layout: "The ground floor of the building is constructed based on the Golden Ratio and the golden number" (Dirección Administración Palacio Barolo).

In The Future...?

This contains the authors' personal thoughts and speculations about the future of architectural design. What is the status quo at the time of this publication? What will geometrically conceived buildings look like in future

Figure 4-36 Palacio Barolo
Source: Wikimedia Commons, CC BY-SA 3.0 (https://creativecommons.org/licenses/by-by-sa/3.0/deed.en).

years? And how will technology affect the process of designing? Let's have a glimpse of some possible scenarios.

Artificial intelligence (AI) has become part of many professions, so it isn't far-fetched to think that it will also affect the work of designers and architects. Thanks to a variety of available software and technologies, it is easier than ever to get inspiration for buildings or components that fulfill certain criteria. Currently, a AI bots can give detailed answers to questions and short prompts.

This is why, from the authors' points of view, it seems like an exciting experiment to carry out some tests about how well these technologies work at the time of this publication. Is it really easy to generate detailed plans of buildings with certain properties, such as including the Golden Ratio? The experiment went through three phases, all differing in the amount of AI used.

Phase 1: Little AI Used

For the first step of testing the ability of AI to generate images of geometrical buildings containing Golden Ratios, the authors used a CAD software called *Rhinoceros 3D* to design a very basic structure consisting of only two Golden cuboids. In the second step, a pre-rendered image of this "building" was implemented into a free-to-use online AI image creator called *Dream* (dream.ai/create). Users prompt *Dream* by describing how the AI should change or manipulate the image. In this case, expressions like "Make it look like a modern building" or "Skyscraper" are practicable suggestions. Some of the final results were astonishingly lifelike, while others were absolutely unrealistic and omitted.

Figure 4-37 shows some of the results of this process. On the left side is the pre-rendered model of the complex, as it was designed in *Rhinoceros 3D*. On the right side are four of the ideas generated by *Dream* for minimalist residential buildings based on the model.

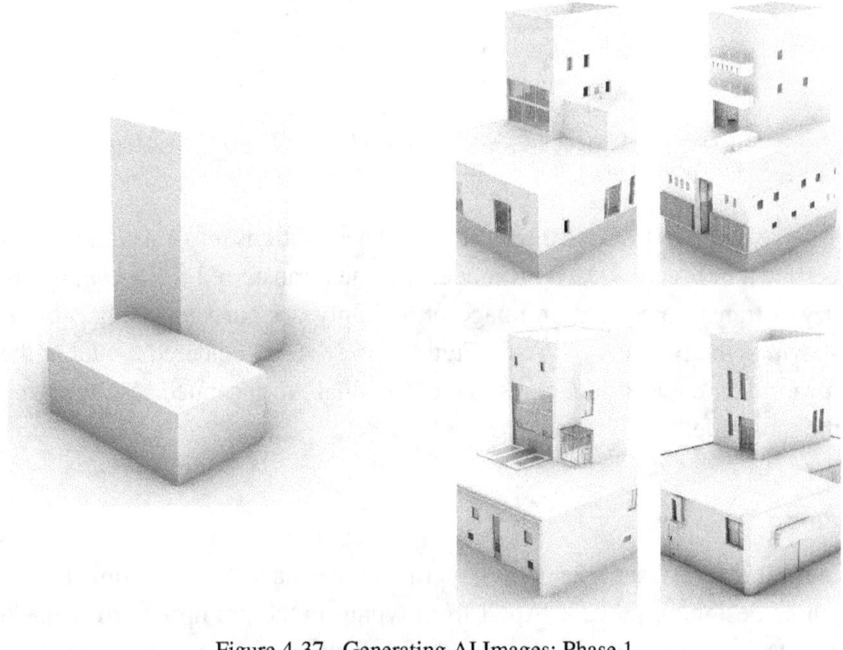

Figure 4-37 Generating AI Images: Phase 1

Phase 2: AI for Design

In the next phase, we sought instructions from AI on how to design the building. In this case, we used *ChatGPT* (openai.com) to get some inspiration for the exterior of a newly designed building, and then followed the same steps as in Phase 1 (modeling in *Rhinoceros* → generating AI images). This program runs on prompts, which the AI uses to generate answers. Here, the task "Design a building with Golden cuboids" was entered.

ChatGPT then gave a detailed verbal answer as to how a building with Golden cuboids might look. As in Phase 1, the AI's answers were not totally logical, and it sometimes contradicted itself, so the complete description could not be transferred into the CAD software. But some other ideas did indeed seem feasible. Here is a short extract from *ChatGPT*'s answer to the prompt:

> *The Golden Cuboid Tower is a modern, multi-purpose skyscraper with a unique and aesthetically pleasing design. The Golden Ratio is applied to determine the dimensions of various elements, resulting in a balanced and visually appealing structure. The tower's height adheres to the Golden Ratio, where the height of the tower is 1.618 times the width of the base. The base of the tower forms a perfect square, and each subsequent level's dimensions follow the Golden Ratio. The Golden Cuboid Tower is a stunning and innovative architectural design that combines the principles of the Golden Ratio with modern construction and design techniques. Its harmonious and balanced appearance is not only visually appealing but also functional for its inhabitants.*

The result of transferring these verbal descriptions into a real 3D model is shown in Figure 4-38. This model was then transferred to *Dream*, where it was transformed into an image of a simply designed residential tower, shown in the Figure 4-38. As in Figure 4-37, the pre-rendered model of the tower is on the left side, and some of the final AI-generated results are on the right side.

Phase 3: Exclusively AI

In the third and last phase of the experiment, we made AI responsible for the whole designing process. Apart from typing in the prompt, no man-made actions were used in the design. To fulfill this task, we used another AI

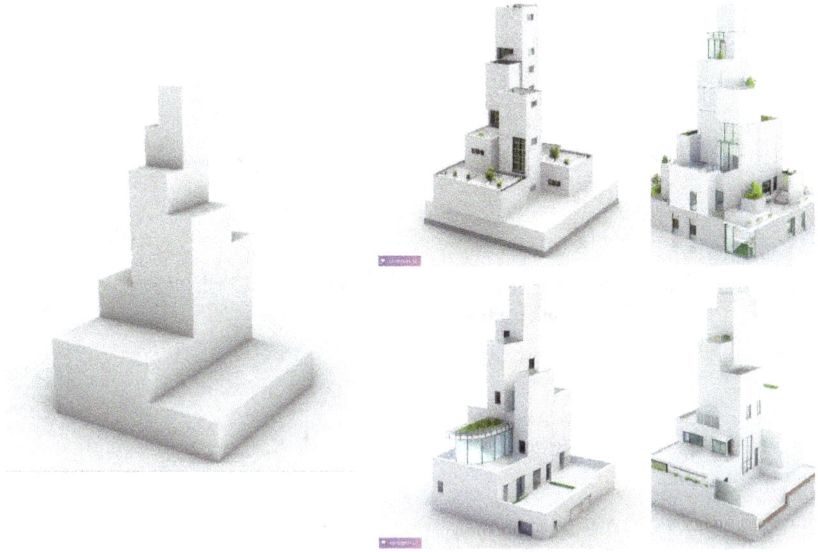

Figure 4-38 Generating AI Images: Phase 2

software called *Bing Chat* (bing.com). This software combines the other two previously used, as it can create both texts (like *ChatGPT*) and images (like *Dream*). The mode of functioning stays the same: Enter a prompt of what the program should do, and it creates four different images within seconds. Some prompts in this case were “Design a modern building with Golden Cuboids” or “Show me architectural sketches of a building containing the Golden Ratio,” etc.

Figure 4-39 shows *Bing Chat*'s architectural sketches of such a building. They look like plans drawn by hand that show the hypothetical nature of these concepts. Figure 4-40, however, depicts four different buildings inspired by the Golden Ratio more realistically. These buildings look like photo-realistic renderings of a complex in its urban context. Both styles are an integral part of planning and designing any architectural structure. In most cases, initial ideas are predominantly captured by hand using pencils, markers, and paper. CAD softwares are then used to model the building virtually render ideas of how the newly planned building will look in context.

Figure 4-39 Generating AI Images: Phase 3 (1)

Summary of the Golden Ratio in AI

None of the ideas shown here claim that any of these concepts will ever be reality. The results of the three phases of the experiment are entirely hypothetical at the time of this publication. First, we cannot predict how architecture will look in the future, nor how much AI will contribute to the development of style. These AI-generated images only show a "snapshot" of contemporary abilities. This is why this experiment used three stages of AI involvement in order to cover a variety of possible effects. But maybe things will develop completely differently in the coming years, decades, and centuries. Second, all the results are just images generated by a simple

Figure 4-40 Generating AI Images: Phase 3 (2)

prompt. They are not designed by a real brain that considers feasibility and practicability. This means that, for the authors, it is impossible to judge if any of the AI-generated concepts could ever be built in reality. They do look lifelike and some of them even look realistic (especially those in Figure 4-37 and Figure 4-38), but the ideas shown in Figure 4-39 and Figure 4-40 seem futuristic and extravagant. The photorealistic aerial views of the complexes in the city are an exciting prediction of how residential buildings may appear going forward, but structures like these may not even be viable.

All the images only show the final exterior design of various buildings. None of them provides any definite plans for how they can be laid out in detail. An architect's work does not only comprise drawing and designing buildings. They also have to make detailed plans of every component, develop feasibility studies, supervise construction sites, and much more. So, AI does not make architects superfluous at this stage—not at all. Hopefully, it will never do so. Designing aesthetically pleasing and practical buildings is not only a generic technical process (at least it shouldn't be), it also depends on the emotions the human designers want to arouse with their works.

Chapter 5

The Aesthetics of the Golden Rectangle in Everyday Objects

Before examining the application of the Golden Ratio and Golden Rectangles in historic and modern everyday objects, it must be made clear that there are mathematical inaccuracies in some objects and proportions may deviate a bit from perfect Golden Ratios. This is because objects like these can't just "look good" to make them more attractive to potential purchasers. Generally, they must also be useful, which influences the object's dimensions and measurements. So, in many cases, the actual dimensions and proportions are determined by aesthetics and usability. In the examples in the following section there are some irregularities due to mathematical inaccuracies in the objects themselves. The presented historic objects were all handmade in ancient times, so they are not perfect in terms of mathematically precise measurement.

In order to understand the measurements of the objects, Romanian mathematician Matila Ghyka (1881–1965) differentiated two types of rectangles:

- "Static" rectangles: These figures have a side-length-ratio of rational fractions, such as $\frac{1}{2}$, $\frac{2}{3}$, or $\frac{6}{19}$.
- "Dynamic" rectangles: These figures have a side-length-ratio of irrational numbers, such as $\sqrt{2}$, $\sqrt{5}$, or ϕ.

In terms of generating interesting, aesthetic structures, dynamic rectangles are more helpful that static rectangles. This is because dynamic rectangles can be geometrically subdivided in different ways to create interesting surface harmonics. Figure 5-1 shows some possible subdivisions of different

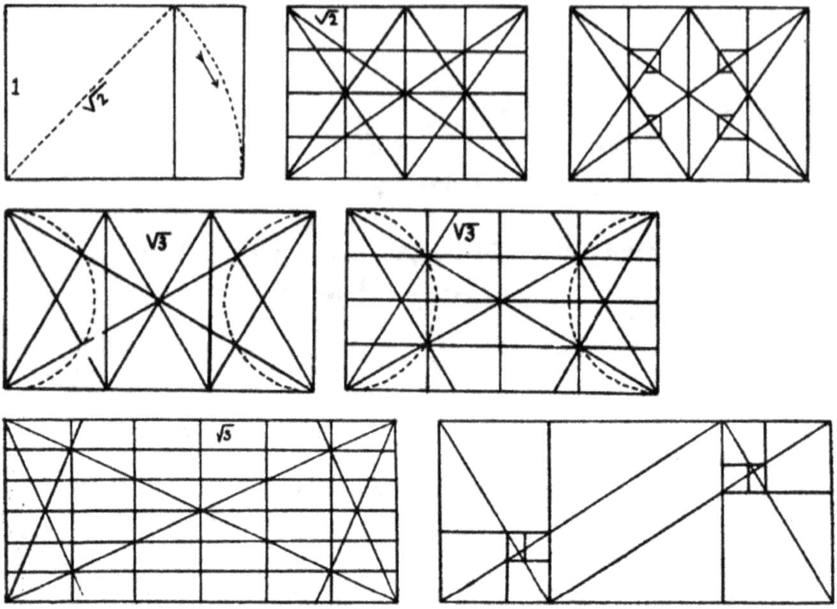

Figure 5-1 Subdividing Various Dynamic Rectangles

dynamic rectangles into smaller parts. The first row shows rectangles with ratio $\sqrt{2}$, the second row of rectangles have ratio $\sqrt{3}$, and the last row shows rectangles with ratio $\sqrt{5}$, which can be seen multiple times in the subdivision of the Parthenon's façade.

The omnipresence of rectangles with side-ratio $\sqrt{5}$—let's call them $\sqrt{5}$-*rectangles* from now on—in the Parthenon's façade is the reason why they play an important role in this section as well.

Historic examples

Greek Vases

Stamnos

The first type of vase described here is called a *stamnos*, a jar with a short neck, a large opening, and small horizontal handles. In this analysis of the stamnos, we discover different ways to subdivide the vase into several $\sqrt{5}$-rectangles or Golden Rectangles. First, the whole vase including its cover and handles is dimensioned to fit exactly two vertical $\sqrt{5}$-rectangles placed side-by-side, shown in Figure 5-2(a). Second, adding a separation line to these rectangles at the height of the lower edge of the neck, as Figure 5-2(b) illustrates, divides both of the $\sqrt{5}$-rectangles into two Golden Rectangles—a small horizontal one and a large vertical one.

Third, when omitting the lid, the vase is contained within three Golden Rectangles: A large one on the lower part reaching up to the handles, and two smaller ones, whose sides-lengths are the side-lengths of the large one shortened by the factor $\frac{1}{2}$, as we can see in Figure 5-2(c). Lastly, the vase can also be geometrically divided following its actual shape more precisely. The handles have to be omitted for that. Without the handles, the vase-body itself can be divided into three Golden Rectangles and two squares. The lid has the dimensions of two squares and two vertical $\sqrt{5}$-rectangles placed side-by-side. Figure 5-2(d) shows the composite of all these shapes.

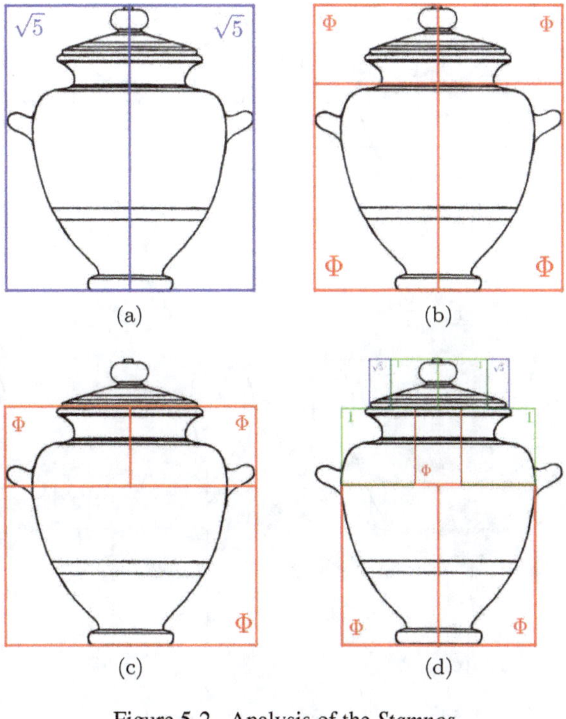

Figure 5-2 Analysis of the *Stamnos*

Kantharos

The second type of vase is a *kantharos*, a chalice-shaped bowl with two large bent handles and a high stem. The whole vase itself can fit almost perfectly into two horizontal $\sqrt{5}$-rectangles stacked on each other, as Figure 5-3(a) illustrates. In terms of Golden Rectangles, the whole vase offers some interesting peculiarities: In Figure 5-3(b), we can see that the width of the vase and the height between the bottom of the stem and the upper edge of the bowl shape a Golden Rectangle. Additionally, the width of the vase and the distance between the upper-most point of the handles and the upper edge of the stem (lower edge of the bowl) shape a Golden Rectangle too. The shaded overlap of these two rectangles measures exactly the height of the bowl. Rotating the two Golden Rectangles into a vertical position and placing them side-by-side (covering the whole vase) generates another overlapping area, which is shaded red in Figure 5-3(c). The small ring connecting the

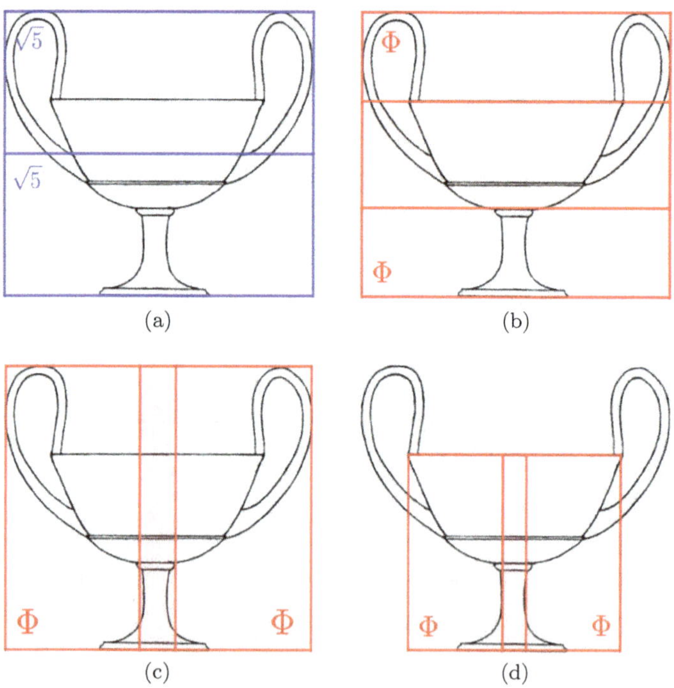

Figure 5-3 Analysis of the *Kantharos*

stem and the bowl fits exactly into this area—they have the same width. The last characteristic pointed out here is another overlapping area of two vertical Golden Rectangles placed side-by-side. This time, their outer vertices are at the outer edge of the bowl, covering only the bowl and the stem, leaving out the handles. Figure 5-3(d) shows that the shaded overlapping area has the same width as the stem at its narrowest part.

Kylix

The last type of vase examined here is a drinking cup called *kylix*. It is characterized by its flat circular bowl and large diameter. There are two handles placed at the sides. The bowl stands on a low stem with a wide base. The first interesting geometrical property of the vase concerns vertical Golden Rectangles, depicted in Figure 5-4(a). We connect the rim of the bowl (omitting the handles) with the middle of the bottom of the stem with a dotted line. This dotted line intersects the lower part of the bowl. The vertical distance between the bottom of the stem and the upper edge of the bowl stands in the Golden Ratio to the horizontal distance between the outmost point of the bowl and the intersection point mentioned before. This is why there is a perfect Golden Rectangle in this space.

Consider horizontal Golden Rectangles: Figure 5-4(b) shows that the bowl of the kylix (omitting the handles and the stem) and the lower edge of the small ring connecting the bowl and the stem are covered by two horizontal Golden Rectangles placed side by side, leaving a small gap in between. This connecting ring is just as wide as the gap. Figure 5-4(c) shows another possible way of separating the vase into simple geometric figures by combining the crucial points shown in Figure 5-4(a) and those in Figure 5-4(b). The intersection of the dotted line and the lower edge of the bowl shown in Figure 5-4(a) and the connection of bowl and stem mentioned in Figure 5-4(b), together with the base of the stem, generate a perfect horizontal $\sqrt{5}$-rectangle.

Furthermore, the height of the *kylix* stands in proportion of $\sqrt{5}$ to the horizontal distance between the vase's vertical middle line and the outmost point of the stem's base. Additionally, the vertical distance between the lower edge of the connection between stem and bowl and the bowl's upper

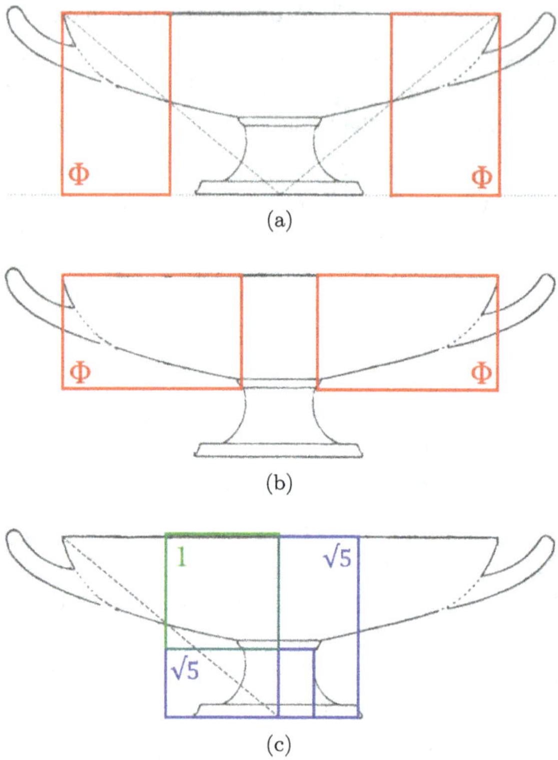

Figure 5-4 Analysis of the *Kylix*

edge is almost identical to the horizontal distance between the connection point from Figure 5-4(a) and the vase's vertical middle line, forming a nearly perfect square.

Other Greek objects

Ritual Pan

Aside from vases, many other handcrafted objects used by the ancient Greeks in their daily life have dimensions following strict geometrical rules. Here we will examine a sacrificial ritual pan. Unfortunately, information about the pan's measurements, material, use, and symbols is not available. This is why we will concentrate exclusively on the proportions of the different parts. Probably the most obvious property of the pan is that the whole object fits perfectly into a Golden Rectangle—the whole length in total is times longer than the whole width (the diameter of the circle), as shown in Figure 5-5(a).

Let's have a more detailed look at the lower part of the pan, depicted in Figure 5-5(b). First, the complete lower part, including the diameter of the circle and the length of the handle, fits inside a Golden Rectangle. Cutting out two vertical Golden Rectangles from the large one (one on each side) leaves some space, which contains the handle. We can further subdivide this space into two identical vertical Golden Rectangles (one at the top and one at the bottom) to generate an overlapping area, shown by bold lines. This area is also a Golden Rectangle. Furthermore, the horizontal lines generated by this subdivision separate the ornaments carved into the handle. Not only is the handle itself dimensioned in a strict geometrical way, but also the decorations follow Golden geometry.

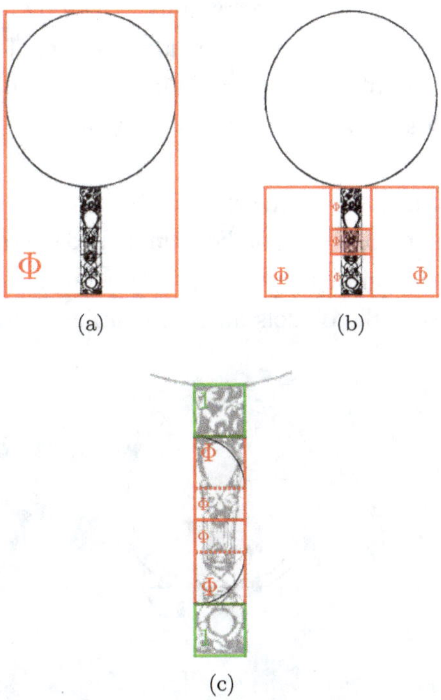

Figure 5-5 Analysis of the Ritual Pan

Lastly, we use Figure 5-5(c) to only analyze the handle (without the geometric context of the pan). The handle is composed of two squares and two vertical Golden Rectangles stacked on top of each other. As already mentioned above, the horizontal lines between these shapes also indicate the different symbols of the ornaments. The newly generated lines also make two smaller horizontal Golden Rectangles when cutting off squares from the initial Golden Rectangles.

Bronze Mirror

An ancient bronze mirror completes the analysis of everyday objects from ancient Greece. It consists of a round plate above a depiction of two winged people facing each other, completed by a long, thin handle. For a deeper look into the geometric proportions of the mirror, some simplifications must be made. The handle is not taken into consideration. The mirror-plate itself is simplified to two circles: a small inner circle representing the actual mirror-face, and a bigger outer circle generated by extending the arc-shaped connection between the plate and the ornament to a full circle. The complicated shape of the ornament of the two winged figures is simplified to an upside-down triangle with a rectangle underneath. Figure 5-6 shows the result of this simplification.

Delving deeper into the geometrical analysis of the mirror, we will first look at the lower part, consisting of the triangle and the rectangle beneath it. This whole complex is close to two vertical Golden Rectangles placed side-by-side. In contrast to the objects analyzed earlier, there are indeed some

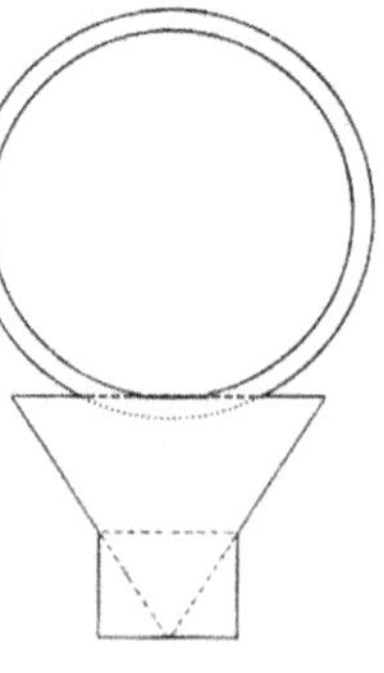

Figure 5-6 Bronze Mirror

bigger deviations from ϕ that are too severe to be ignored. Figure 5-7(a) shows the situation, where the side ratio of the two rectangles is approximately 1.726. To determine the relative deviation from ϕ, we use the formula of *relative difference*: $d_r = \frac{d_a}{v_{ref}}$, where d_a is the absolute difference (here $1.726 - \phi$), and v_{ref} is the reference value (here ϕ). This gets us $\frac{1.726-\phi}{\phi}$, which is 0.067, meaning that the two rectangles differ from perfect Golden Rectangles by about 6.7% each. Due to these inaccuracies, ϕ is written only in brackets in the diagram. A similar situation occurs when looking exclusively at the small bottom rectangle of the simplified structure of the mirror: It also consists of two single vertical rectangles, whose proportions are only close to ϕ, as can be seen in Figure 5-7(b). This is why ϕ is written there in brackets. The divergence from a real Golden Rectangle comes to about 9.3%. Starting with these rectangles, we now construct a square above them, so that its upper edge touches the outer circle of the mirror-face. Starting from its top left vertex, we will construct another square in such a

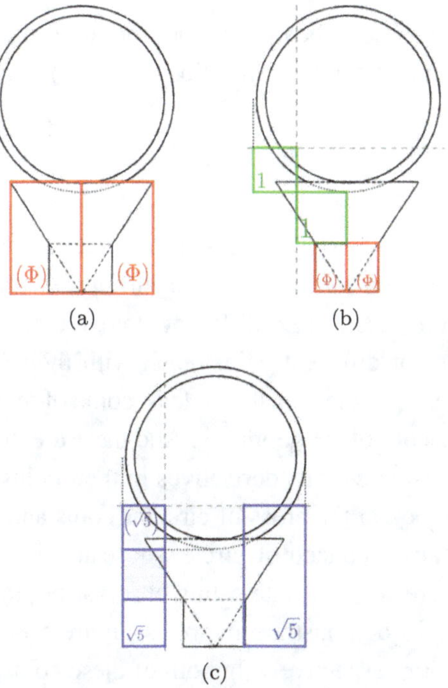

Figure 5-7 Analysis of the Bronze Mirror

way that its left edge aligns with the edge of the outer circle. The top and the right sides of this new square mark straight lines that are important for the next step of the analysis, which is shown in Figure 5-7(c).

For this step, $\sqrt{5}$-rectangles are elaborated. Between the straight lines generated in Figure 5-7(b), the left edge of the pan, and the upper edge of the rectangle at the bottom of the mirror fits a rectangle, whose proportions are close to $\sqrt{5}$. They differ by about 3.2%. Between the vertical gray straight line, the lowest point of the rectangle at the bottom of the mirror, and the left and the lower edges of the outer circle fits a perfect $\sqrt{5}$-rectangle. On the right, there is also another accurate $\sqrt{5}$-rectangle being placed into the space between the outmost edge of the pan, the horizontal gray straight line, and the outer edges of the rectangle at the bottom of the handle.

Modern and Contemporary Examples

Physical Objects

Does the ancient Greek dedication to incorporating the Golden Ratio into ordinary objects continue today? How does ϕ influence the process of design these days?

Coins

There is an immense variety of coins in circulation all around the world. Nearly every country on earth has its own currency (with the biggest exception being the Euro, used in 26 different states) and, thus, its own coins. The sheer number of different coins, each with their own measurements and motifs, may prompt the idea that at least some of them are dimensioned following the rules of golden geometry. And there are indeed signs of some major currencies using ϕ or its derivatives in their coins.

First, we will look at the areas of circular coins and their relation to ϕ. When a coin contains a concentric circle whose area is ϕ times smaller than the whole coin, there are some interesting observations to be made. The bold circles in Figure 5-8 demonstrate them, as Figure 5-8(a) and (b) depict a 1- and a 2-euro coin, respectively. In both of these coins, the circle touches the twelve stars arranged in a circular arc around the coin's inner part. These

Figure 5-8 The ϕ and the Area of Some Coins

stars are a well-known symbol of Europe and also a crucial element of the European Union's flag.

Figure 5-8(c) and (d) show coins of the Japanese yen (¥). The golden coin in Figure 5-8(c) has a value of ¥5 and shows a circular arrangement of Japanese *kanji* characters. The upper ones say *Nippon-koku*, meaning *Country Japan*—the state's official title. The lower ones indicate the year of the minting. We can see that the *kanji*'s outer edges touch the red circle—they all fit exactly inside. Figure 5-8(c) shows a copper-colored coin of ¥10. It has a circular wreath of leaves enclosing the coin's numerical value and the year of the minting. This wreath's circumference exactly matches the red circle, "growing" inside of it.

The three coins shown in Figure 5-9 approach ϕ differently. The yellow circles within these coins have radii ϕ times smaller than the radii of their respective coins. Figure 5-9(a) shows the ¥50 coin. Just as in Figure 5-8(c), the circle determines the placement of the circularly arranged *kanji* telling the year of the minting, but in this case the characters are outside the circle.

(a) (b) (c)

Figure 5-9 The ϕ and the Radius of Some Coins

Returning to European currencies, Figure 5-9(b) shows a coin of 5 Swiss francs. In this case, the yellow circle exactly encloses the head of the figure; specifically, it encloses the man's hood. Figure 5-9(c) depicts a coin of 1 Icelandic króna, showing its value in words, the state's name in Icelandic, and the year of minting. The yellow circle here exactly encloses the figurative motif of the coin—a depiction of a giant. On the coin, the figure is actually surrounded by a visible circle, which is precisely congruent to the yellow circle. This is why it can barely be seen in Figure 5-9(c). The giant is an ancient legendary symbol of Iceland, and it is also part of the country's coat of arms.

Banknotes

The same principle of the many different coins all around the world can also be applied to banknotes. There are many different banknotes circulating worldwide, and chances are that at least some of them make use of the Golden Ratio ϕ. One type of rectangle including ϕ is the Golden Rectangle with an additional $\sqrt{5}$-rectangle beside it, as shown in Figure 5-10. So, when

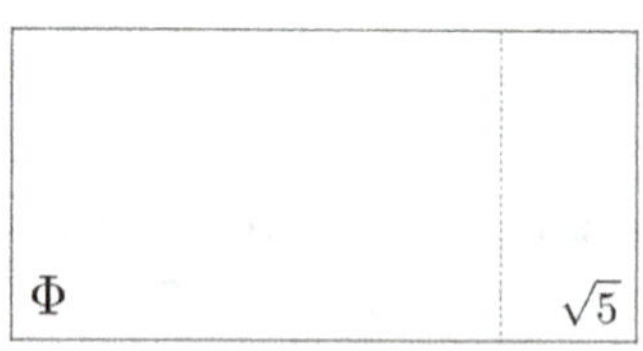

Figure 5-10 The $\phi + \frac{1}{\sqrt{5}}$ Rectangle

standardizing the shorter side to length 1, the longer side has length $\phi + \frac{1}{\sqrt{5}} = \frac{7+\sqrt{5}}{2\cdot\sqrt{5}} \approx 2.065$.

The banknotes of various currencies use "$\phi + \frac{1}{\sqrt{5}}$ rectangles" in their designs. One example is the Swiss franc, whose 100-franc note has measurements of 144 mm × 70 mm. The ratio of its width to its length is $\frac{144}{70} \approx 2.057$. The relative deviation from the note to the rectangle mentioned above is about 0.4%. Even closer to the ratio of $\phi + \frac{1}{\sqrt{5}} = 1.618 + .447 = 2.058$ is the Icelandic 500-króna note. This bill measures 145 mm × 70 mm, making a width-to-length ratio of $\frac{145}{70} \approx 2.071$. This number only differs from $\phi + \frac{1}{\sqrt{5}}$ by about 0.3%, but this is still not the ideal. Let us consider the 1-yuan bill of the renminbi, the People's Republic of China's currency. This note has dimensions of 130 mm × 63 mm, and its sides ratio of $\frac{130}{63} \approx 2.063$ shows a relative deviation of only about 0.11%. In other words, if its width were only 0.05 mm shorter, the 1-yuan banknote would be a (nearly) perfect $\phi + \frac{1}{\sqrt{5}}$ rectangle. Figure 5-11 shows the two banknotes presented here. Their deviations from a $\phi + \frac{1}{\sqrt{5}}$-rectangle cannot be seen with the naked eye.

Figure 5-11 Banknotes

Violins

When discussing musical instruments, it is hard to determine their dimensions just in terms of aesthetics. After all, their primary purpose is to produce clear tones and sounds. The dimensions of the individual components of an instrument are chosen to optimize the quality of the tones and make the instrument as comfortable to play as possible. Aesthetics, therefore, play a subordinate role. Yet there are some proportions in violins that suggest ϕ. The violin shown in Figure 5-12 is 56.6 cm long. Probably the most prominent separation line is the one dividing the violin into an "upper" part—consisting of the neck, the pegbox, and the scroll—and a "lower" part—consisting of the body. This segment divides the whole violin in nearly exactly the Golden Ratio. The upper part is 22.3 cm long, the lower part is 34.3 cm long: $\frac{55.6}{34.3} \approx 1.621 \approx \phi$. The deviation to ϕ is only 0.18%. Furthermore, the distance between the pegs and the lower ribs is divided in nearly the Golden Ratio by the same line as the previous example: The distance between the separation line and the pegs measures 12.6 cm, the one between the separation line and the ribs measures 20.34 cm. Thus, there is a ratio of $\frac{20.34}{12.6} \approx 1.614$. It deviates from ϕ by about 0.23%. Additionally, the distance between the neck's lower edge and the f-hole's lower edge is 8.61 cm, and from there to the bow of the tailpiece is 5.32 cm: $\frac{8.61}{5.32} \approx 1.6184$. This ratio differs from ϕ only in the fourth decimal. Numerically speaking, the deviation is only about 0.02%.

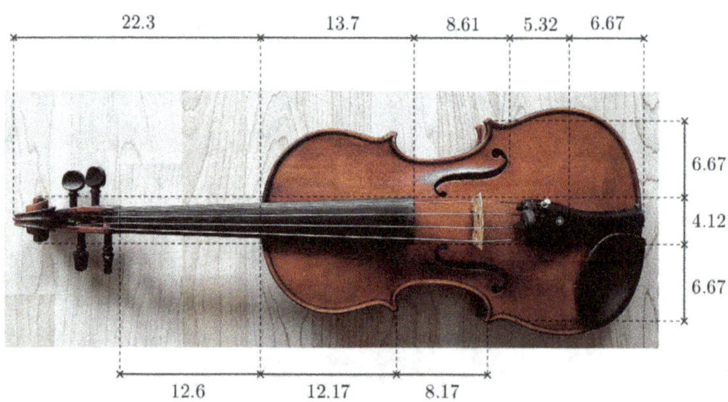

Figure 5-12 Measurements of a Violin

As previously mentioned, the whole upper part of the instrument measures 22.3 cm. The ratio of this length to the distance of 8.61 cm between the bottom of the neck and the bottom of the f-hole from the example above yields $\frac{22.3}{8.61} \approx 2.59.\phi^2$, however, is 2.618. Thus, there is a difference of this ratio to ϕ^2 is only 1.07%.

All the examples stated above refer to the length of the violin's components. But what about the width of the violin? The width of the instrument between the lower ribs is 17.46 cm. The scroll, the neck, and the tailpiece all measure 4.12 cm in width, which means that the width between the neck and the ribs measure 6.67 cm each. Their ratio is $\frac{6.67}{4.12} \approx 1.619$, a value that differs from ϕ by only about 0.06%.

Logos and Icons

In this section we will consider different well-known logos to see whether the designers employed the Golden Ratio. Are these icons also designed following the old idea of ϕ? Or does this number no longer play an important role in modern design?

iCloud

Apple's online cloud service, iCloud, has changed its logo several times. One of the previous logos looked very similar to that shown in Figure 5-13.

The logo depicts a stylized cloud. It is composed of four circular elements, each with an outer and an inner edge generating a white outline. Do their dimensions follow a pattern that is in accord with ϕ? None of the circles stand in a perfect Golden Ratio to another in any regard. We cannot find exact ϕ here. But some of them do approach it. A closer look at the arcs

Figure 5-13 iCloud Logo

comprising the outline reveals that they are *not* perfect circular arcs, but are slightly irregular. However, these deviations are small enough to be ignored here, which is why the components are still called "circles" or "circular arcs" and are also treated that way, even though it is, strictly speaking, not correct.

First, we will consider the measurements of the inner circles, depicted in Figure 5-14.

All these circles are now examined in terms of the proportions of both their radii and their areas. Figure 5-15(a) shows the two circles whose radii are in a relation that is closest to ϕ. The icon does not have a fixed universally applicable size, as it is used in different contexts and different screens that depict it in different sizes. We cannot set the radii to fixed dimensions. Thus, we can only standardize the smaller circle's radius to 1 for this case, which makes the bigger circle's radius measure about 1.67. We use the formula of relative difference $d_r = \frac{d_a}{v_{ref}}$, where d_a is the absolute difference $(1.67 - \phi)$, and v_{ref} is the reference value (ϕ), to get $\frac{1.67-\phi}{\phi}$, which is about 0.0321, meaning a 3.21% difference from ϕ. Figure 5-15(b), however, gives those

Figure 5-14 Geometrical Analysis of the iCloud Logo (1)

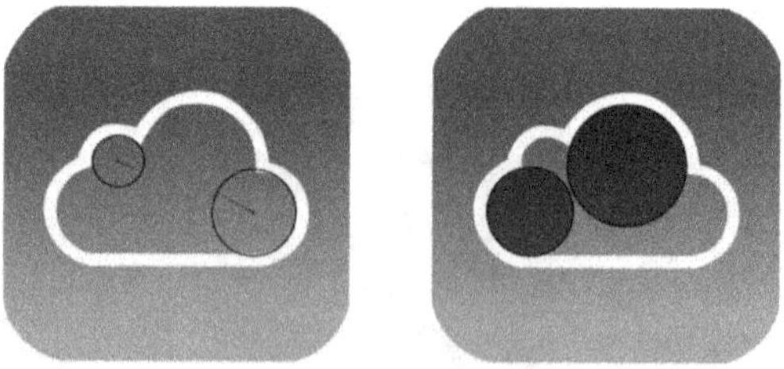

(a) (b)

Figure 5-15 Geometrical Analysis of the iCloud Logo (2)

two circles, whose areas are in a relation that is closest to ϕ. When we standardize the smaller circle's area to 1, the bigger one has an area of about 1.898. We see a way bigger deviation from ϕ here: Using the same formula as above, we get a result of about 17.3 %.

We will now repeat all the calculation steps done with the smaller circles with the outer, bigger ones, shown in Figure 5-16.

Figure 5-17(a) gives those two circles whose radii represent ϕ as closely as possible. When the smaller radius has length 1, then the bigger one has length 1.845. Using the formula of relative difference again, we calculate $(1.845-\phi)/\phi \approx 0.1403$, or 14.03%. Looking at the areas of the circles in the same way, those two in Figure 5-17(b) approximate ϕ best. Standardizing the smaller circle's area to 1, the bigger one has an area of about 1.695. The relative difference in this case is $(1.695 - \phi)/\phi$, or about 4.76%.

So far, we have compared the small and the large circles separately. When analyzing both small and large circles comprehensively, there are some proportions to be found which are even closer to ϕ. Again, Figure 5-18(a)

Figure 5-16 Geometrical Analysis of the iCloud Logo (3)

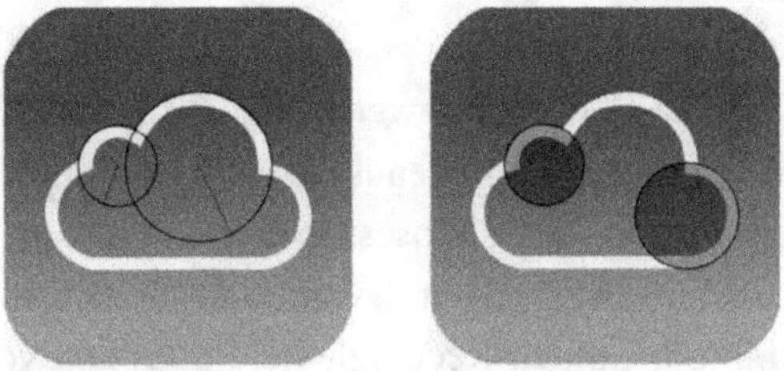

(a) (b)

Figure 5-17 Geometrical Analysis of the iCloud Logo (4)

(a) (b)

Figure 5-18 Geometrical Analysis of the iCloud Logo (5)

shows the one pair of circles whose ratio of radii is closest to ϕ. With a smaller radius of 1, the bigger radius measures 1.68. Here, the formula of relative difference gives a result of $(1.68 - \phi)/\phi \approx 3.83\%$. And now, as the last aspect, the areas: Figure 5-18(b) shows those two circles whose areas reflect ϕ best. The smaller circle has an area of 1, and the bigger one has an area of 1.643. Again, when using the same formula, we get a result of a deviation from ϕ of about 1.54%.

All in all, the old iCloud Logo does indeed have some geometrical shapes that relate to ϕ quite closely. The best approximation to that number has a difference of about 1.54%. But still, there is not a single proportion in all the circles that meets the Golden Ratio exactly. And some of the proportions deviate too far from ϕ. To say that this number was, generally speaking, the inspiration for the design of circles in the logo would be a stretch. But there still is indeed a perfect Golden Ratio integrated into the icon—not in the circles, but in the whole cloud itself. The outermost points of the inner edge of the cloud-outline fit perfectly into a Golden Rectangle, as Figure 5-19 shows.

The reason why we analyzed a logo which isn't in use anymore is that the current iCloud logo harbors even fewer golden proportions. It now consists of only four circles (not eight) because there is no outer and inner edge anymore. The proportion that is closest to ϕ in the new logo is the ratio of the areas of two circles that show a deviation of about 2.04%. Also, fitting the whole cloud into a Golden Rectangle here does not work as accurately as in the previous logo in Figure 5-19.

Figure 5-19 Golden Rectangle in the iCloud Logo

Car Brand Logos

Today's automobile market is a fiercely contested economic sector and offers a large variety of different brands to potential buyers, all with their own advantages, disadvantages, and unique selling points. To make a certain car brand recognizable in this maze of different brands, a unique logo plays an important role. The logo is associated with the brand itself and should be easy to remember and attractive to the owner or potential purchaser. This is the reason why many well-known brands use the Golden Ratio somewhere in their emblems. The ones presented here only mark the tip of the iceberg—there are many more examples for the use of golden geometry in car brand logos.

Figure 5-20(a) shows the emblem of the Japanese company Toyota. It is composed of three ellipses that shape a stylized T inside an oval. Let's consider the width of the whole logo (called a) and the horizontal distance between the emblem's outer edge to the outer edge of the vertical central ellipse (called b). These two lengths are in the Golden Ratio: $\frac{a}{b} = \phi$. Furthermore, the logo's total height (called c) stands in the Golden Ratio to the vertical distance between the logo's bottom and the inner edge of the horizontal upper ellipse (called d): $\frac{c}{d} = \phi$. The logo of Mazda—another Japanese brand—can be seen in Figure 5-20(b). The emblem, composed of a stylized M in an oval, fits perfectly into two vertically-oriented, side-by-side Golden Rectangles. Additionally, the separation line generated by cutting off squares from the bottom of the rectangles is at nearly the same height as the intersection of the "diagonals" of the M and the surrounding oval.

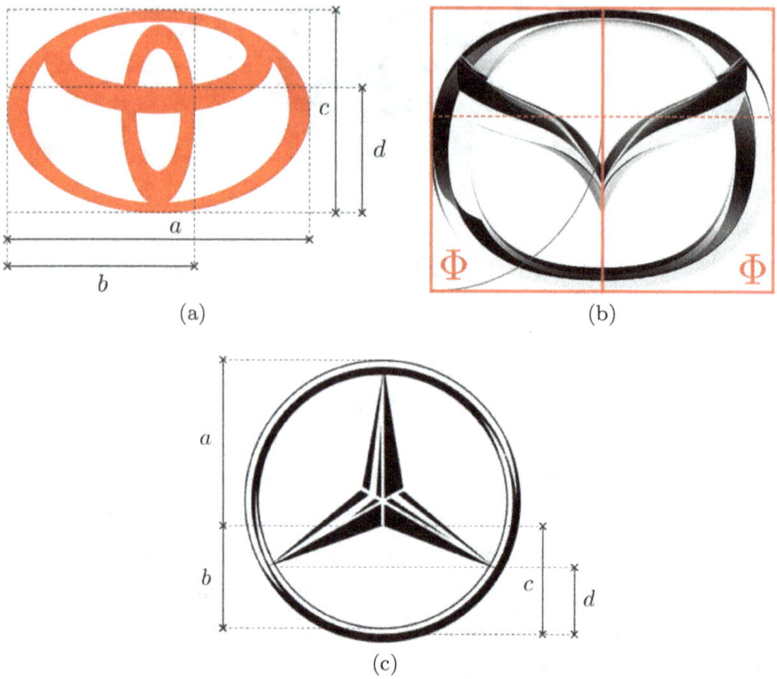

Figure 5-20 Golden Ratios in Car-Brand Logos

The last example of a car logo containing the Golden Ratio is the German brand Mercedes-Benz and its famous three-pointed star, shown in Figure 5-20(c). The circle circumscribing the star consists of an outer and an inner ring. Let the distance between the outer ring and the point where two rays diverge be called a, and the distance between the same point and the inner ring be called b. Just these lengths show the Golden Ratio: $\frac{a}{b} = \phi$. Another incidence of ϕ can be discovered in the emblem, where a circle in the middle of the ring creates a 3D effect. The distance between this ring and the previously mentioned ray-separation point (called c) and the distance between this ring and the outer tips of the rays (called d) are in the Golden Ratio, too: $\frac{c}{d} = \phi$.

Chapter 6

Geometry in Nature

Symmetry

Whether it is in a human face, a leaf, or in birds, symmetry affects the world in which we live. It is the basic organizational framework of human nature and our environment. But what is symmetry? Many people have the basic idea that an object is symmetrical if, when divided in half by a straight line or a plane, two identical halves are created. In mathematics, this is an insufficient description. Mathematics sees symmetry in terms of movements: A symmetrical object can be moved without its appearance being changed. In other words, the object looks the same after the movement as it did before the movement. To describe symmetry mathematically, one can use group theory.

Symmetry groups are special forms of mathematical groups. There are three essential criteria for symmetry groups. The first characteristic is the existence of a *neutral element*, which is often denoted by e. In the multiplication of integers, for example, this would be 1, which does not change the initial element. For example: $a \cdot 1 = a$.

In the concept of symmetry, there is a neutral element named *trivial symmetry*. If one applies this to an image, then the image is unchanged. In general: $a \cdot e = a$. The second requirement for groups is the existence of an *inverse element*. In the case of symmetry, this means that every symmetry also has an opposite symmetry. Using multiplication as an example, we could choose 7 as any element of the set of integers. The inverse of 7 is $\frac{1}{7}$. When the two elements are combined, that is, multiplied, the number 1 is the result. In other words, we have $7 \cdot \frac{1}{7} = 1$. Thus, if any element of the set is combined with its inverse, the neutral element is obtained. In general: $a \cdot a^{-1} = e$. The third criterion of symmetry groups is *concatenation*. If one executes a symmetry two or more times, a symmetry of the image is created again. For instance, if we rotate an image by 90 degrees twice in succession, we end up with a 180-degree rotation. This means that a new symmetry operation is produced by repeating an existing symmetry action.

If all three criteria — the existence of a neutral and inverse element as well as the concatenation — are fulfilled, one can speak of a symmetric group. Symmetry is, therefore, quite easy to define mathematically based on a few criteria. Moreover, it is also a constant part of human existence. Everywhere we go, there appears to be symmetry. But technically, that is not true. Why is that? Nature is not perfect. There are small "errors" like scratched edges and marks, which, strictly speaking, make the bodies asymmetrical. Thus, in the real world, it would be hard to find exact symmetry. And yet, people idealize symmetry. Humans associate it with beauty and use it to structure their environment. There are different ways of doing this. If you ask a biologist, they will cite two types of symmetry, namely, bilateral and radial symmetry. If a beetle's wings are described as bilaterally symmetrical, it means that a cut in the middle would produce two mirror-image identical halves (Figure 6-1). This type of symmetry can also be seen in flowers. The human body also exhibits bilaterally symmetrical on the outside, although our organs aren't all symmetrically arranged.

In contrast to bilateral symmetry is radial symmetry. Starting from the center, circular areas of the same appearance grow outwards. It can be seen in starfish as well as other plants and animals, as shown in Figure 6-2.

Figure 6-1 Examples of Bilateral Symmetry in Animals and Plants
Source: Alfred Schrock, United States Geological Survey, Unsplash.

Figure 6-2 Examples for Radial Symmetry in Animals and Plants
Source: Indraneel Pole, Phillip Larking, Mink Mingle, Unsplash.

Figure 6-3 Horn Trefoils Demonstrate Bilateral and Radial Symmetry Simultaneously
Source: Stan Slade, Unsplash.

Considering symmetry, one can divide the animal kingdom into the Radiata and Bilateria. To the former belong jellyfish, starfish, radiolarians, and more. The larger part, however, is taken by the Bilateria, to which all other classes of animals and plants belong, such as birds, humans, insects, etc. There are also creatures that have a hybrid of the two symmetries, such as the horn trefoil shown in Figure 6-3. The inflorescences are arranged radially, that is, they form outward in a circular shape. The flowers themselves, on the other hand, have a simple bilateral symmetry.

Tessellations

Plane tessellation, or tiling, is an arrangement of closed shapes that completely fill a surface without leaving gaps. We find natural tessellations

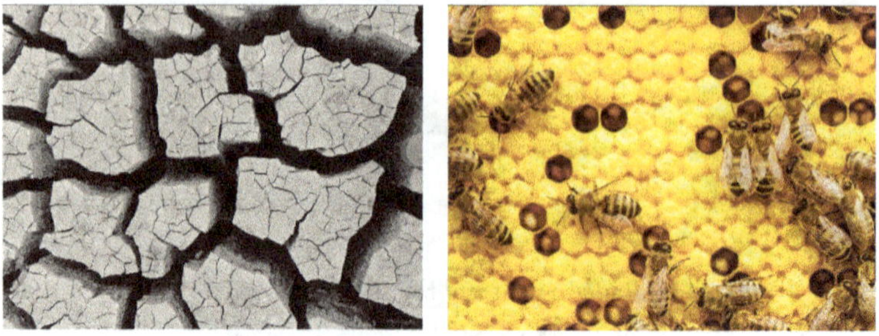

Figure 6-4 Examples of Tessellation in Nature
Source: Joashua Brown; Simon Kadula, Unsplash.

in honeycombs and the mud of dried lakes, as shown in Figure 6-4, as well as artificial tessellation in architecture and other areas of our everyday life.

Honeycombs

We are familiar with the honeycomb pattern exhibiting its regular hexagons. The German mathematician Johannes Kepler (1571–1630) claimed that bees possess a mathematical understanding because of how they develop these hexagons. He had no other way to explain the accuracy of these living creatures. But why, exactly, do bees use hexagons for their structures at all? Alternatively, they could form squares or circles. To answer this question, it is critical to understand what purpose the honeycombs serve for their bees. On the one hand, the honeycomb serves as a birthing and living space for the bees' larvae. On the other hand, the insects store honey and pollen inside the honeycomb. The bees build the walls with wax, which they produce themselves by sweating out the mass with their wax glands. The builder bees knead the wax with their mouthparts so that it becomes soft. This process is extremely strenuous for the construction bees, which is why they want to save as much energy as possible when building. The bees' goal is to be able to raise as many larvae as possible in as small a space as possible. Let's consider which construction method is the best way for the builder bees to achieve this goal by comparing different types of packaging methods. The larvae are approximately circular in their initial stages. As a first idea, we start setting the larvae in a square arrangement, as shown in Figure 6-5.

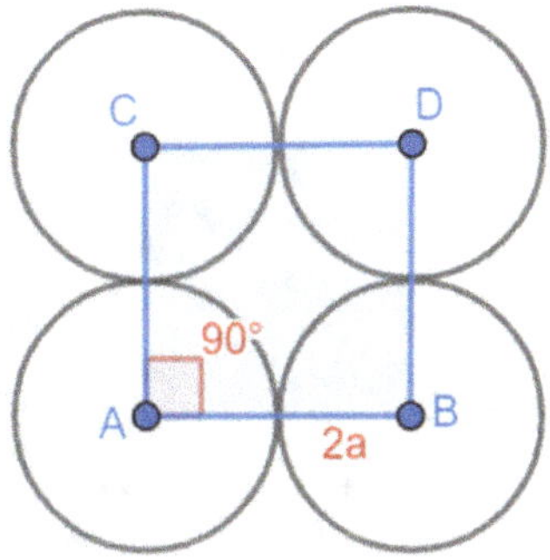

Figure 6-5 Square Packing Arrangement

Let the side of the square be $2a$, and it has an area of $4a^2$. The square partially covers four circles, which represent the larvae. The covered sectors of the circle have a radius of length a. The sum of their areas is equal to a whole circle, which is πa^2. We will now examine the level of effectiveness. This refers to the area that is actually occupied by the larvae in relation to the total area that is used by the square arrangement:

$$\frac{\text{area covered by larvae}}{\text{area of square}} = \frac{\pi a^2}{4a^2} = \frac{\pi}{4} \approx 0.7835981\ldots.$$

According to the calculation, more than 78% of the square's area is covered by larvae. Although this is already quite a high percentage, the bees are not yet satisfied with this. This is because temperature control, in addition to making the most use of available space and resources, is crucial for creating honeycombs. For the larvae, it is vital to maintain a constant temperature of about 34°–36° Celcius. Although the bees can raise the temperature by trembling their wings, this process consumes a lot of energy, which reduces the effectiveness level of the square arrangement. The square construction method does not seem to be the most suitable.

We now consider improving efficiency to better protect the larvae. Let's consider the shape that the builder bees have actually chosen. In Figure 6-6, we can see that the hexagonal arrangement, which is made up of six triangular segments, can be more efficient. Let's determine the efficiency of a triangular arrangement and a hexagonal arrangement. Again, the side length of the equilateral triangle is $2a$. The area of the equilateral triangle is $a^2\sqrt{3}$.

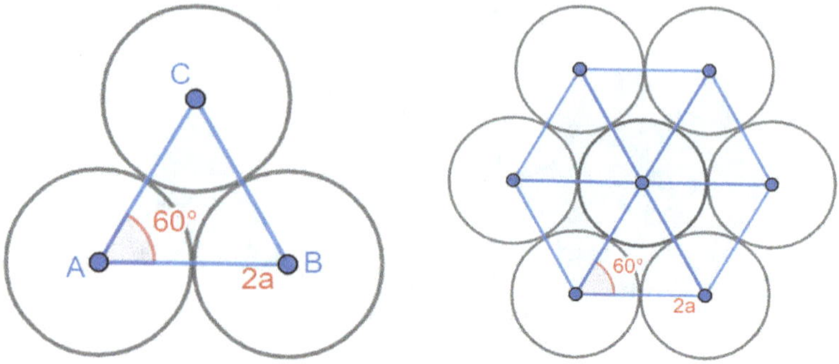

Figure 6-6 Hexagonal Packing Arrangement

The triangle covers three sectors of a circle with radius a, and the area of the covered sectors is equal to a semicircle, $\frac{\pi a^2}{2}$. Therefore,

$$\frac{\text{area covered by larvae}}{\text{area of triangle}} = \frac{\frac{a^2\pi}{2}}{a^2\sqrt{3}} = \frac{\pi}{2\sqrt{3}} \approx 0.9068996.$$

as shown in Figure 6-6.

Therefore, a hexagonal arrangement based on equilateral triangles covers more than 90% of the circular larvae. So, the bees are justified in using hexagonal honeycombs, which can be built much denser than square honeycombs. The bees save energy and material and do not have to worry about temperature regulation as much. Of course, one could test other polygons or other shapes to prove that the hexagon is the densest packing for circular surfaces. But this is no longer necessary thanks to the French mathematician Joseph Lagrange (1736–1813). Although his work did not specifically focus on the densest arrangement of circular surfaces, he still drew great conclusions about it. Lagrange's two main statements, formulated in 1773, were as follows: First, there is no ordering structure of circles denser than the hexagonal arrangement. And second, only the hexagonal order provides the highest density. Bees, of course, did not have to wait until Lagrange's insight to build the most efficient arrangement. The bees evolved the building process to save energy and materials. Nevertheless, to ensure stability, they built the honeycombs with regular hexagons. Bees aren't the only ones;

Figure 6-7. Alhambra and its Tessellations
Source: Kadir Celep, Unsplash.

humans also use tessellations, the overlap-free, area-filling arrangement of shapes. We find other examples of tiling in architecture, such as at the Alhambra, the castle complex in Granada, Spain, which is one of the most visited attractions in Europe. In 1984, it became a UNESCO World Heritage site. The exterior and interior walls are covered with intricate colorful mosaic patterns, such as the one shown in Figure 6-7.

Footballs

Even in sports there is no avoiding tessellations. The soccer ball is a wonderful example of an overlap-free arrangement of partial surfaces. Prior to 1962, the soccer ball appeared different from today's style and consisted of 12 surfaces. But that changed with the advent of color television. Not only was the number of surfaces increased to 32, but also the colors were changed for better contrast with the green turf. Nowadays, the soccer ball consists of black pentagons and white hexagons. Thus, the soccer ball is composed of polygons as shown in Figure 6-8. This guarantees that there are no gaps and no overlaps. For it to take on its spherical shape, the ball must be inflated with high pressure.

Figure 6-8 Soccer Ball as a Tessellation

Circles

In nature, we rarely find straight lines. In fact, the most natural curve in our living world is the circle. If we were to walk across the desert without a navigation system, we would not walk straight ahead, but in a circle. This is partly because our left stride length is not equal to our right stride length. Another explanation is that we cannot estimate the position of the sun without error and, thus, we could lose our orientation. The circles we are walking are called *path curves*. This means that the circle consists of a combination of locomotion and an angle.

Such curves are not only seen in desert migration, but also in leafcutter bees. These insects eat leaves, which they cut out in an approximately circular shape. The beetle proceeds as follows: It cuts out very short stretches of equal length with its mouth. The lines form an angle of slightly less than 90° in the direction in which the leafcutter is moving. The result is a regular polygon that looks like a circle because of its many vertices.

You can define circles not only as path curves, but also as distance curves. This means that all points have the same distance to the center. We can see these kinds of curves in the form of small waves, for example, when a stone is thrown into water. When the stone falls into the water, the liquid must move out of the way. The stone displaces the water, and the water particles move up and down. As they do so, they regularly disperse outward, creating concentric circles as shown in Figure 6-9.

Figure 6-9 Concentric Circles Formed in Water
Source: Nick Fewings, Unsplash.

Trees

Many plants grow by gradually building up layers externally. If this process is uniform, that is, linear, circular rings are formed. In the case of trees, these rings can even be used to determine the age of the tree. From year to year, starting from its center, the tree trunk creates new layers annually. In spring, a light and thick layer is formed. In summer and autumn, the darker, thinner parts form. Together they form an annual ring. If you count fifty such rings, for example, the tree would be fifty years old (see Figure 6-10). In tropical trees, the annual rings are less pronounced. Due to the higher temperature, the wood grows more evenly and forms thinner layers overall.

Not only is the inside of a tree exciting to observe geometrically, but the tree's external growth also fascinates mathematicians. Trees can measure several dozen meters in height. At the same time, they are exposed to various environmental factors, such as strong gusts of wind and rain. To withstand storms, they develop a strong root system that is firmly anchored in the ground. The transition between the underground parts and the aboveground trunk of the tree is not straight, but also curved. This reduces the tension in the wood and thus avoids predetermined breaking points. To determine the ideal curve shape for the growth of trees, the German physicist Claus Mattheck (1947–) developed the *Method of Tensile Triangles.*

Figure 6-10 Linear Growth in a Tree Trunk

Source: Joel & Jasmin Førestbird, Unsplash.

The first tensile triangle is constructed with two straight lines that are perpendicular to each other. These represent the roots and the trunk of the tree. In the next step, a triangle is constructed by connecting the root and the trunk at an angle of 45° (see Figure 6-11). To build the second triangle, bisect the hypotenuse of the first triangle. This length forms the radius of a semi-circle centered at the upper vertex of the first triangle. If you connect the intersection of the trunk and the semi-circle with the center of the

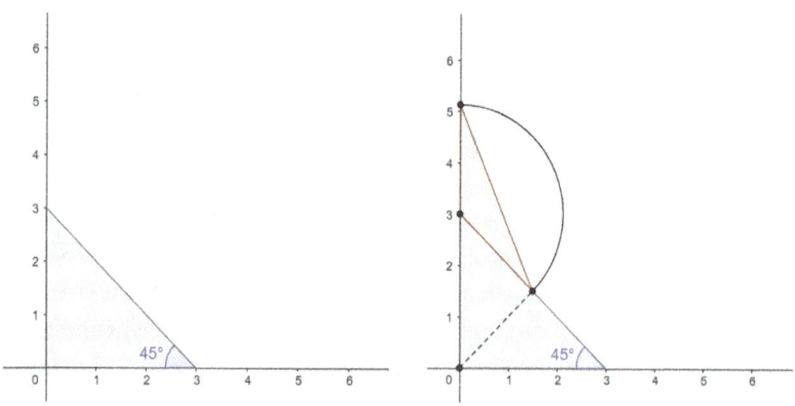

Figure 6-11 Construction Steps of Tensile Triangles

Figure 6-12 Method of Tensile Triangles on a Tree Trunk
Source: Mak, Unsplash.

midpoint of the triangle's hypotenuse, you get the second triangle. The third triangle is created in the same way as the second one (see Figure 6-12). If the method of tension triangles is applied to the design of components, they should withstand a breaking test about ten times longer than components without this design.

Spirals

There is no way to avoid the Golden Ratio and the Fibonacci numbers, especially in the world of plants. If you count the petals of a plant, the result is often a Fibonacci number (1, 2, 3, 5, 8, 13, 21, 34, 55, 89, ...) as shown in Figure 6-13. Roses often have 34 or 55 petals, two Fibonacci numbers.

Certain plants have spiral-shaped seed arrangements. You quickly see that, when counting the spirals, in most cases the number corresponds to a Fibonacci number here as well. Biologists refer to this unique arrangement of seeds, leaves, and flowers as *phyllotaxis*. Examples of it include sunflowers or pinecones, as shown in Figure 6-14.

Plants, therefore, exhibit the Fibonacci numbers and the Golden Ratio and make use of them. In this section, we will take a closer look at the sunflower. The arrangement of its seeds has evolved over thousands of years. It is crucial for the flower to produce as many seeds as it can since it uses them to reproduce. The more seeds the plant has available for birds,

Figure 6-13 Fibonacci Numbers in Petals
Source: Karl-Heinz Müller; Tim Mossholder; Ikhsan Fauzi; Tobias Rademacher. Unsplash.

the more likely it is that the seeds will spread and germinate. One might think that, like bees, the sunflower uses a hexagonal honeycomb pattern to arrange its seeds, as this is a particularly efficient use of space. However, this would require each seed to be placed according to its final size, that is, without growing directly next to another seed to prevent overcrowding. This goes against the way seeds naturally grow, which is why the sunflower, as with many other plants, has developed a different strategy: The plant produces its first seeds near its center, where they are still small and tightly packed together. As the sunflower develops and the seeds grow, the seeds push outward at an angle from the center of the sunflower and separate themselves from other seeds. To prevent the seeds from lying on top of each

Figure 6-14 Examples for Seed Arrangements
Source: Jirasin Yossri, Jennifer Burk, Unsplash.

Figure 6-15 Parastichies in the Sunflower

other, the distance to the center is also increased with each seed. The aim is to keep the increase in radial distance as small as possible to make the most efficient use of the available space. The angle of twist of the seeds remains almost constant at approximately 137.51°, which corresponds to the Golden Angle. The distribution of the grains at this particular angle creates a multitude of left and right spirals, also known as parastichies, visible to the human eye, which can be seen in Figure 6-15.

Fractals: Self-Similar Structures

Consider coastlines, crystals, and blood flow. These and many more things in our world seem to have nothing in common at first glance, but they are connected by a crucial quality called self-similarity. The French-American mathematician Benoît Mandelbrot (1924–2010) was the first and, for a long time, the only scientist to study self-similar structures. In 1974, he gave the naturally occurring and artificially produced patterns the name "fractals," derived from the Latin word *frangere*, which means "to break." This name describes the typical structure and dimension of fractals. Some of the curves cannot be described as one-dimensional or as two-dimensional objects, as their dimension is "broken" and lies mostly between one and two. An essential quality of fractals is self-similarity and scale invariance. But what does that actually mean? Whatever scale you choose, you can enlarge the image as often as you like, and you will always see the same patterns emerging (see Figure 6-16).

Figure 6-16. Self-Similarity in Romanesco Broccoli
Source: Venus Major, Unsplash.

The Koch Curve

The Koch curve is arguably the most famous fractal object. In this case, the curves are characterized by being continuous everywhere, but differentiable nowhere. The construction of such a curve is iterative. As shown in Figure 6-17, at Step 0 the construction consists of a simple section *a*, which we call the initiator. At Step 1, we make a reduced copy that shortens the line

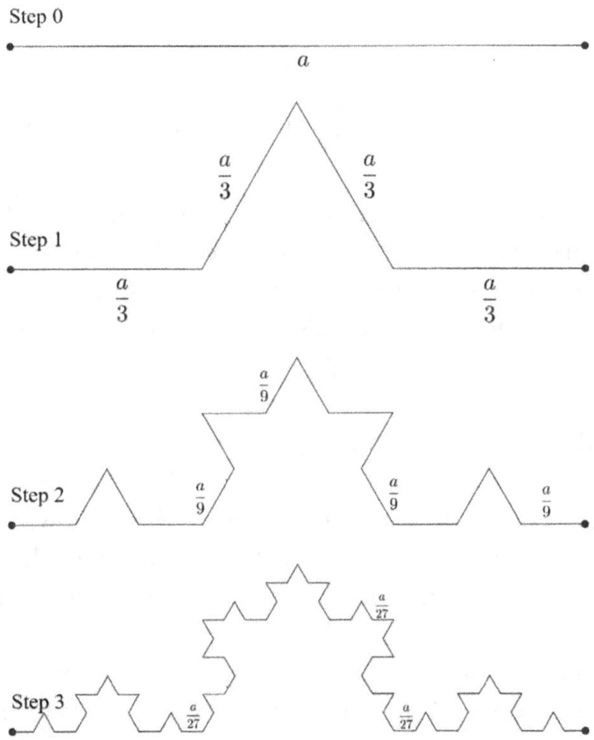

Figure 6-17 Construction of the Koch Curve

length to $\frac{a}{3}$. Four of the shortened distances are lined up as shown above, so that the angles between the individual distances are 240°, 60°, and 240°. If you create a copy of this generator by scaling the structure by a factor of $\frac{1}{3}$ and replacing the simple lines again, you get the image on Step 2. The lines of Step 2 are now $\frac{a}{9}$ long. This process can be repeated as often as desired.

If we now look at the total length of the curve, this can be expressed by formulas as follows:

$$L_0 = \left(\frac{4}{3}\right)^0 \cdot a$$

$$L_1 = L_0 \cdot \frac{4}{3} = \left(\frac{4}{3}\right)^1 \cdot a$$

$$L_2 = L_1 \cdot \frac{4}{3} = \left(\frac{4}{3}\right)^2 \cdot a$$

$$L_3 = L_2 \cdot \frac{4}{3} = \left(\frac{4}{3}\right)^3 \cdot a$$

$$L_n = L_{n-1} \cdot \frac{4}{3} = \left(\frac{4}{3}\right)^n \cdot a \forall n \in \mathbb{N}_0.$$

For this we name the length of our curves L and look at n arbitrary steps. The first length L_0 is made up of a simple section of length a, as described above. It is multiplied by $(\frac{4}{3})^0 = 1$ for illustration of the following iteration. The length L_1 corresponds to four partial stages of length $\frac{a}{3}$, so it can also be said to have length $L_0 \cdot \frac{4}{3}$. Then L_2 is calculated in the same way. Starting from length L_1, the distance is first divided into thirds. Thus, three partial distances of length $L_1 \cdot \frac{1}{3}$ are obtained. Four of these partial distances replace the simple distance by arranging them triangularly and obtain the total length of $L_2 = L_1 \cdot \frac{4}{3}$. Repeating this process n times results in the general formula $L_n = L_{n-1} \cdot \frac{4}{3}$. We can determine the length L of the limit curve, that is, the entire Koch curve, by using the limit value: $L := \lim_{n\to\infty} L_n = \lim_{n\to\infty}(\frac{4}{3})^n \cdot a = \infty$. Thus, the Koch curve has an infinite length. The known curve satisfies the criterion of fractals mentioned earlier as it is (exactly) self-similar. So, no matter how many times you zoom in, you always get the original structure of the object.

The Koch snowflake, which is created by combining three Koch curves, does not fulfill this criterion. It is only mirror-, point-, and rotation-symmetrical, as shown in Figure 6-18. And because we mentioned dimensions at the beginning of this chapter, the snowflake has a fractal dimension of 1.26, while the Koch curve has a dimension of 1.

The Hilbert Curve

Actually, it's not that simple to locate curves that completely fill a two-dimensional space. But in 1891, German mathematician David Hilbert (1862–1943) explored just such curves. These so-called Hilbert curves increasingly fill the plane with each iteration step (Figure 6-19). Corals, among other creature, exploit this quality when populating reefs. They try to occupy as much surface area as possible in a limited space to

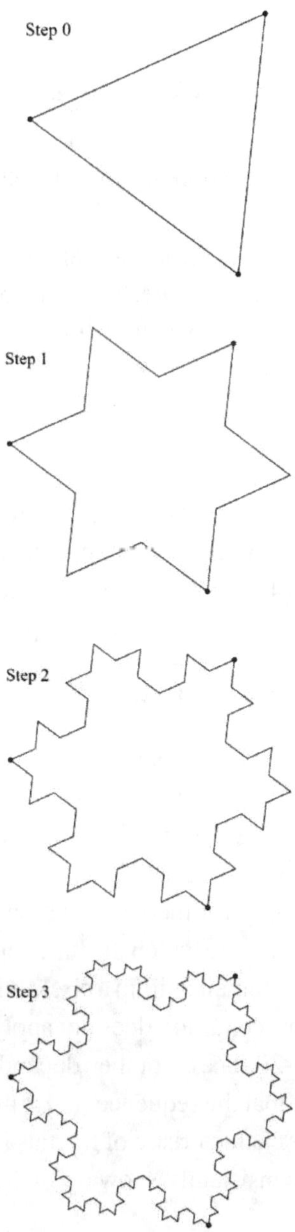

Figure 6-18 Construction of the Koch Snowflake

Figure 6-19 Construction Steps of a Hilbert Curve

increase their food intake. Soft corals can capture plankton and solids more efficiently due to their planar structure. The same principle is found in roots, which can reach nutrients at the bottom more easily due to their strong branching.

The Mandelbrot Set

Before the introduction of graphics-capable computers in the early 1980s, the world of fractals was relatively unknown. That technology made it possible to view previously unknown fractals, such as the Mandelbrot set. This set can be described by a sequence of complex numbers $z_0, z_1, z_2, \ldots$.

$$z_0 = 0, \quad z_1 = f(z_0) = c, \quad f(z_1) = c^2 + c, \ldots, \quad f(z_n) = z_n^2 + c.$$

Mandelbrot sets, therefore, correspond to the set of all complex numbers c for which the complex sequence $\{c, c^2 + c, (c^2 + c)^2 + c, \ldots\}$ with the initial value $z_0 = 0$ is bounded. To put it in other words, their limit value is not infinite ($\neq \infty$).

This means the following for the visualization of Mandelbrot sets: You start with a complex number c, to which the function $f(z) = z^2 + c$ is applied several times. If the results remain within a circle with a radius of 2, the value c is marked in black. If this condition does not apply, c is colored differently, as can be seen in Figure 6-20. The color then depends on how many iterations are required to determine that the sequence $\{c, c^2+c, (c^2+c)^2+c, \ldots\}$ does not remain bounded. The characteristic of fractals is also met in this instance since the same patterns consistently show up in the image regardless of the scale at which it is seen.

We sometimes recognize comparable images in marble stone. The veins in marble can show fractal structures that have a similar structure to the Mandelbrot set, as shown in Figure 6-21.

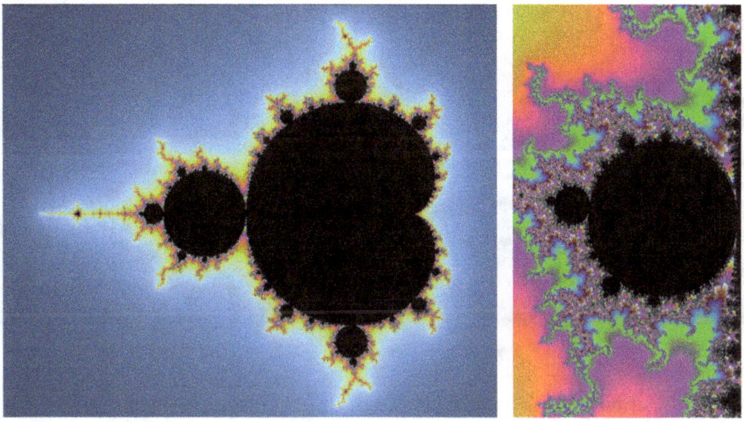

Figure 6-20 Left: A Mandelbrot Surface. Right: The Same Mandelbrot Surface, Magnified (Zoom Factor: 16,000,000)

Source: With permission of Prof. Dr. Edmund Weitz.

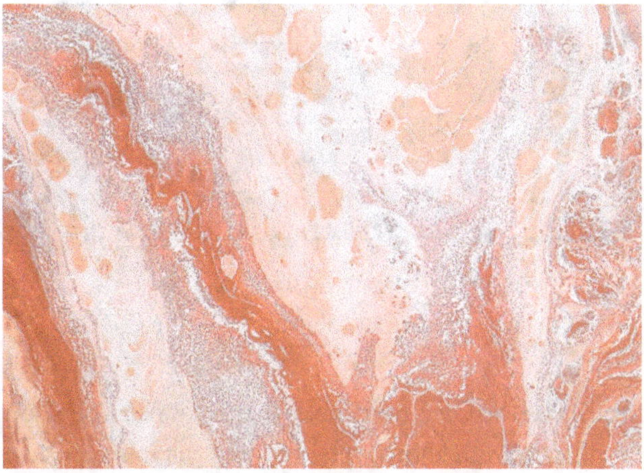

Figure 6-21 Marble as a Self-Similar Structure

Source: Pawel Czerwinski, Unsplash.

Julia Sets

Julia sets, named after the French mathematician Gaston Julia (1893–1978), can be defined with the same sequence as the Mandelbrot sets, with the difference that here the starting point does not have to be $z_0 = 0$ but can be

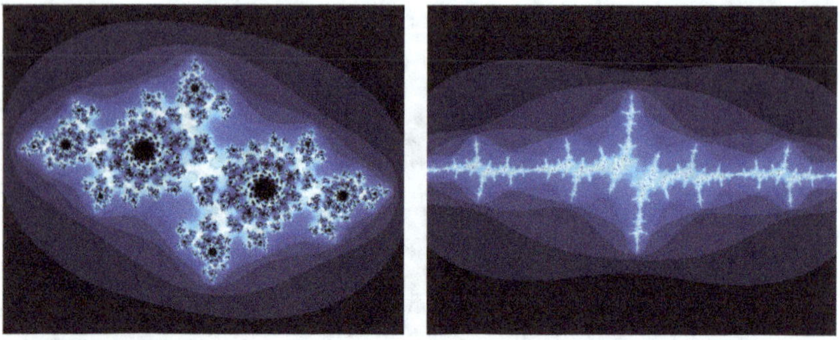

Figure 6-22 Julia Sets

Source: With permission of Matthias Wesker.

Figure 6-23 Large Fires and Clouds as Ever-Changing Shapes

Source: Chirag Nayak; Anna Spencer, Unsplash.

chosen arbitrarily on the Gaussian plane. In their appearance, some Julia sets may seem similar to Mandelbrot sets, but they can also look very different depending on how the value c is chosen. In Figure 6-22, we can see a Julia set with the value $c = -0.77 + 0.3i$ (i being the imaginary unit, $\sqrt{-1}$) in the left-hand image. The right image shows a Julia set with the value $c = -1.4 + 0.05i$.

We can also find this structure in our environment, for example, in large fires and clouds, such as those shown in Figure 6-23. Although they do not follow any precise mathematical formulas, they change their appearance all the time and never settle for an exact shape, just like the Julia sets.

A Summary of Fractals

One aspect we have already established is that fractals can be seen a lot in nature and our everyday life. Even if not everyone knows them by name, fractals are everywhere. Their geometric structure helps us to understand the biological structure of nature better. And the better we understand nature, the more thoughtfully we can deal with it. The study of fractals can be used for species protection and medicine. One example is the study of the heartbeat. The central organ pumps in a fractal pattern, which can be monitored in cardiology. If the heart beats too irregularly, this may indicate congestion and subsequent heart failure. If, on the other hand, the pattern is too irregular, experts may conclude that the heart is in fibrillation. By studying the heartbeat more closely, it should be possible to determine early diagnoses and thus save lives.

The circulatory system also takes advantage of its fractal structure. Its self-similarity is shown by repeated branching, which resembles a tree. Why these developed along evolution is easy to understand: Blood wants to get to as many places as possible. Nevertheless, the human body should not consist only of blood vessels, which is why it is also important for it to take up as little volume as possible. Arteries and veins can increase their surface area by branching out, thus creating a blood network that helps them supply blood to as many points in the body as possible.

It is worth noting that, beyond our natural world, fractal geometry found its place in the film industry around the 1970s. The American computer graphics researcher Loren Carpenter (1947–), after studying Mandelbrot's work, developed several concepts to use fractals in computer graphics. He focused mainly on the design of mountain landscapes and finally published "Vol Libre," the first computer-animated short film, in 1980. Although "Vol Libre" may not be as impressive as modern animation, Carpenter's film had a great impact on the movie world. The short, a moving representation of artificially created mountains, was a sensation at the time. The fractal geometry behind it is still used today, especially in science fiction. Graphic designers can create a wide variety of visual effects with fractals, such as planets or explosions.

Chapter 7

Conic Sections

Historical Overview

Conic sections have a long history and date back to the ancient Greek scholars who studied their geometric properties centuries ago. Over time, outstanding mathematicians such as the Greek mathematicians Hippocrates of Chios (470–410 BCE), Menaichmos (380–320 BCE), Apollonius of Perga (240–190 BCE) and the German mathematician Johannes Kepler (1571–1630) have made significant contributions to the research and classification of conic sections. Their findings and classification systems are still used in mathematics today. The approach to conic sections is inextricably linked to the classical problem of cube duplication, which has challenged generations of mathematicians.

Cube Duplication

The world of mathematics is full of fascinating riddles. One of them is cube duplication, which (together with squaring the circle and angle trisection) is one of the three classical problems of ancient mathematics. In fact, cube duplication is not only a mathematical challenge, but also the cause of a highly interesting Greek legend named the Delian problem. According to the legend, the inhabitants of the island of Delos suffered a devastating plague. They asked their oracle for advice, and it set the inhabitants the difficult task of doubling the volume of the cube-shaped altar that was standing in the magnificent temple of Apollo.

Mathematicians of the time had to find the side length of a cube with twice the volume using only compasses and rulers. Many tried in vain to solve this problem—a problem that still leaves mathematicians in wonder. But what is it that makes this task so difficult? If we consider a cube with side length a, its volume is $V_1 = a^3$. The side length of the cube with twice the volume is called x, resulting in the volume $V_2 = x^3$. Thus, we are looking for the unknown length x of the equation $x^3 = 2 \cdot a^3$. To solve the equation for x, take the third root of both sides. This gives the expression $x = a\sqrt[3]{2}$. Figure 7-1 illustrates the task: The purple cube with the side length $x = \sqrt[3]{2}$ has twice the volume of the green unit cube with $a = 1$.

If the side a is given, one can immediately calculate the unknown length $x = a\sqrt[3]{2}$ in a numerical way. But how should the edge length x be determined geometrically? It was not until the 19th century that the French mathematician Évariste Galois (1811–1831) proved that the Delian problem cannot be solved graphically. The core of the proof is based on the fact that the irrational number $a\sqrt[3]{2}$ cannot be expressed by integers, square roots, or the four basic arithmetic operations.

The Greek mathematician Hippocrates of Chios made a significant contribution to the cube duplication problem by transforming it into another one using two mean proportions.[1] Thus, the Delian problem, where the side length a is described by $x^3 = 2 \cdot a^3$, is solved if it is possible to

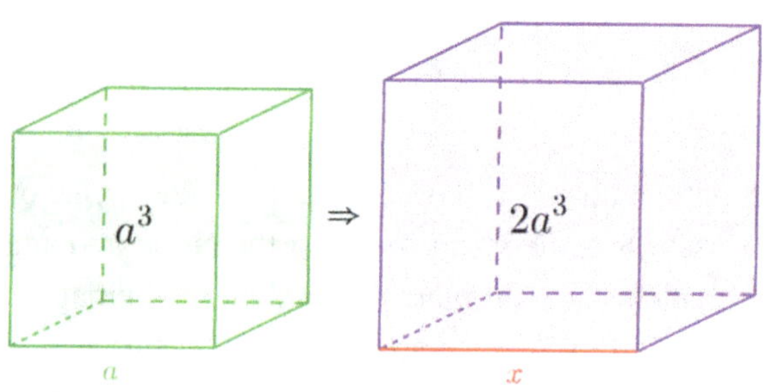

Figure 7-1 Cube Duplication

[1] Meaning that two ratios are set equal to each other and change in the same ratio to each other.

determine another length y such that the following condition $\frac{a}{x} = \frac{x}{y} = \frac{y}{2a}$ applies.[2]

Years passed, and nobody was able to solve the oracle's riddle. But one mathematician looked at the problem from a different perspective and (eventually) found a solution. Menaichmos considered the proportions separately and transformed them into two individual equations. He then discovered two conic section lines, which are now known as parabola and hyperbola. From $\frac{a}{x} = \frac{x}{y}$, one gets the equation of a parabola $y = \frac{x^2}{a}$, the green curve shown in Figure 7-2. From $\frac{a}{x} = \frac{y}{2a}$, one gets the equation of a hyperbola $y = \frac{2a^2}{x}$, the purple curve shown in Figure 7-2. At the intersection coordinate $S(x, y)$, where the conic section lines that are obtained from the equations intersect, the x-coordinate corresponds the side length a of the original cube and the y-coordinate corresponds the sought side length x of the cube with double volume.

As stated, the purple cube with the side length $x = \sqrt[3]{2}$ has twice the volume of the green unit cube with the length $a = 1$. This is also demonstrated

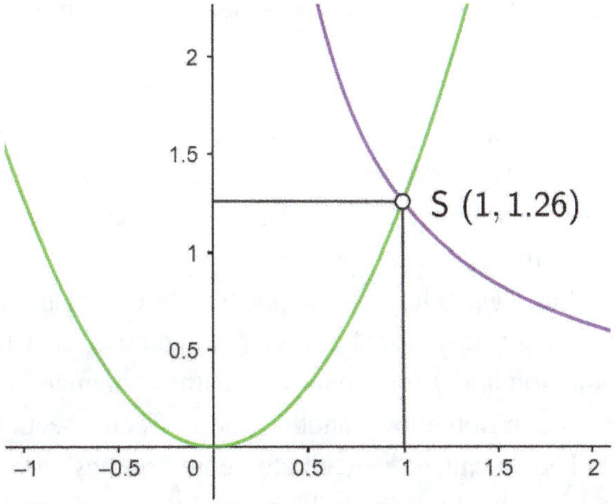

Figure 7-2 Curves after Menaichmos

[2] In words: The variable a stands for the side length of the original cube. It is the given variable that represents the initial volume of the cube. The variable x is the desired side length of the cube with twice its volume. The variable y is another variable describing the relationship between a and x.

in Figure 7-2, in which the y-coordinate of the intersection point S of the two curves provides exactly the required side length $\sqrt[3]{2} \approx 1.26$ for doubling the volume based on the unit cube with side length 1.

Cone Sections are Sections

Apollonius of Perga (240–190 BCE) was an important Greek mathematician who worked intensively on conic sections. In his renowned work *Conica*, he not only described basic definitions and properties of the ellipse, parabola, and hyperbola, but also gave the conic sections their corresponding names. In addition, Apollonius proved that the three different intersection curves originate from the same general type of cone: the double cone of revolution. His innovative methods and concepts left a lasting impact on mathematics and influenced subsequent researchers such as Johannes Kepler and the English mathematician Isaac Newton (1643–1727). Around two millennia later, the French mathematician Germinal Pierre Dandelin (1794–1847) discovered the most elegant proof of the conic sections. In order to get an idea of Dandelin's proof, a visual representation of the intersection curves is required.

Figures 7-3, 7-4, and 7-5 illustrate four different intersections of a double circular cone with planes. The double circular cone is created when a straight line is rotated around an intersecting axis. The cone's lateral surface consists of the totality of all straight lines, which are called generatrixes. If the lateral surface of the double cone is now intersected with a plane, intersection curves are created. Depending on the position of this plane, four different shapes are obtained: an ellipse, a circle, a parabola, or a hyperbola. In mathematical jargon, those four shapes are named "regular conic sections." But in fact, the definition allows another form of conic sections that is not regular at all. The so-called "degenerate conic sections" are obtained if a section plane runs exactly through the apex of the cone, resulting in either a point, a pair of straight lines, or a straight line.

An ellipse, shown in Figure 7-3, is created if the section plane is not parallel to any generatrixes. This means that the inclination angle β of the plane is smaller than the opening angle α of the double cone. If this angle β is zero, a circle—the special case of an ellipse—occurs.

Contrary to the ellipse, a parabola, shown in Figure 7-4, is generated if the intersection plane is parallel to exactly one generatrix of the double

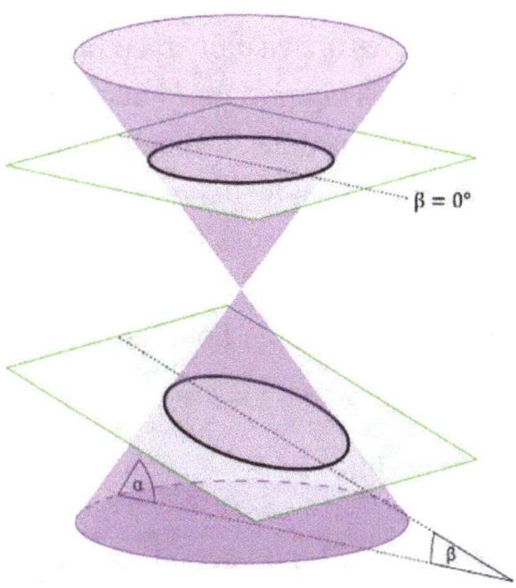

Figure 7-3 Circle & Ellipse

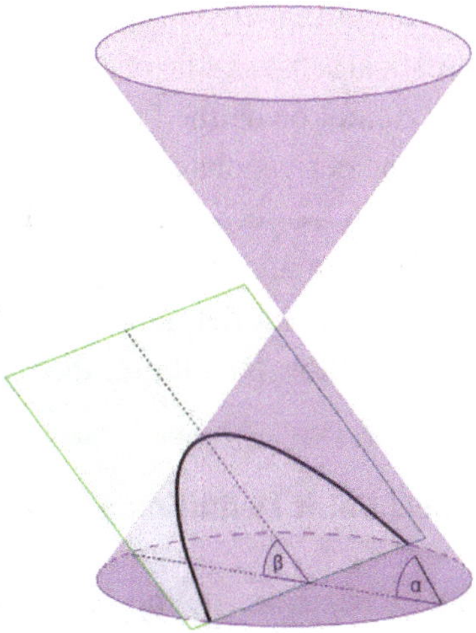

Figure 7-4 Parabola

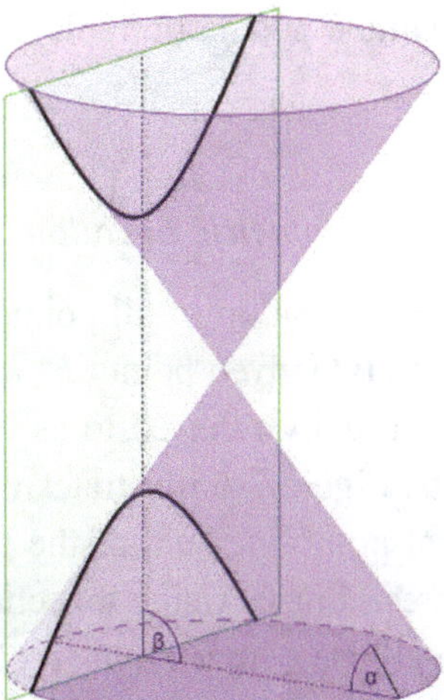

Figure 7-5 Hyperbola

cone. For this, the inclination angle β of the plane must be equal to the opening angle α of the double cone.

Finally, a hyperbola, shown in Figure 7-5, is obtained if the section plane is parallel to two generatrixes of the double cone. This means that the angle β of the plane is greater than the opening angle α of the double cone.

Unfortunately, this way of defining conic sections is not ideal in a mathematical sense, as the resulting intersection curves are visually recognizable but cannot be defined precisely. Nevertheless, it is important to understand that the derivations and definitions of conic sections must ultimately correspond to the sectional plane that is shown. We will now delve deeper into the mathematics to verify the fit of the conic section lines by introducing the focal point definition and Dandelin's spheres. We will also reveal the secret of how the ellipse, the parabola, and the hyperbola got their names.

Focal Point Definition and String Construction

The focal point definition of conic sections is a fundamental geometric concept in which ellipses, hyperbolas, and parabolas are characterized by the relationship between the distances to the focal points and a constant quantity. This definition allows for the precise description and construction of conic sections using simple geometric tools such as a string, ruler, and pencil.

The Ellipse—Defined as a Geometric Location

The ellipse is the geometric location of all points P of a plane for which the sum of the distances to two given points F_1 and F_2 is equal to a given constant. The constant is usually referred to as $2a$ and the points F_1 and F_2 are called focal points. Figure 7-6 illustrates these terms. According to the definition of the focal point, the sum of the two purple lines must be twice the length of the green line a. Using this characteristic property, it is possible to construct the ellipse with a simple tool.

The String Construction of the Ellipse

String has been used to construct conic sections since the time of the mathematician Apollonius. This special way of constructing an ellipse is also called "gardener's construction" and works as illustrated in Figure 7-7. The gardener positions two pegs in the ground, exactly at the two foci F_1 and F_2. He then attaches a string of length $2a$ to the foci and guides a rod along the taut string. The tip of the rod describes an arc of the ellipse.

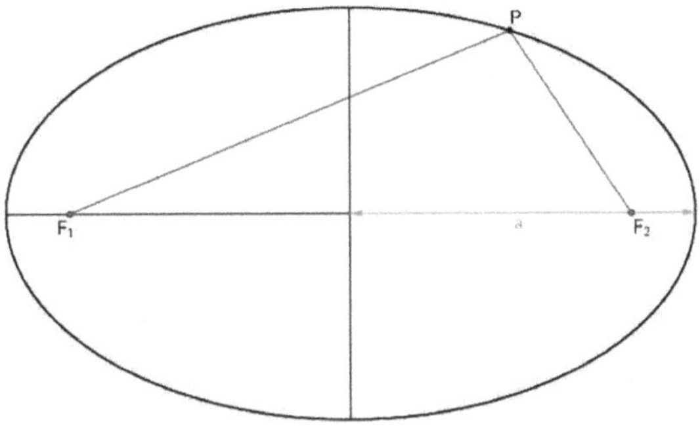

Figure 7-6 Focal Point Definition of an Ellipse

Figure 7-7 Gardener's Construction of an Ellipse

The Hyperbola—Defined as a Geometric Location

A hyperbola is the geometric location of all points P in a plane for which the absolute value[3] of the difference between the distances to two fixed focal points F_1 and F_2 is the same as $2a$. Figure 7-8 shows that the difference between the two purple distances is twice the length of the green distance a. Thus, the hyperbola can be constructed by using simple tools as follows.

[3]The absolute value of a number indicates the distance between the number and zero. Since distances are always positive or zero, amounts are also always positive or zero.

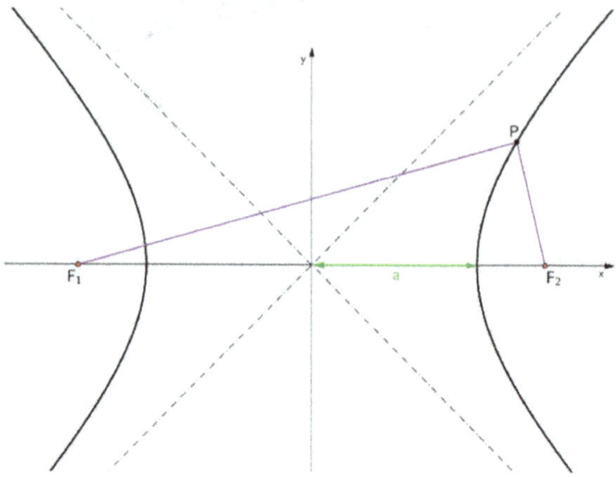

Figure 7-8 Focal Point Definition of a Hyperbola

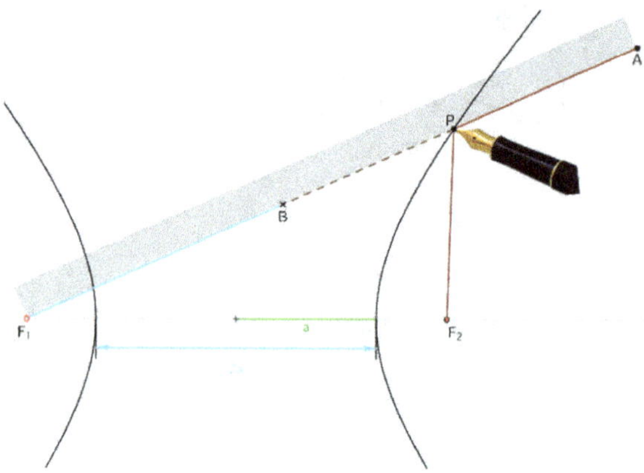

Figure 7-9 String Construction of a Hyperbola

The String Construction of the Hyperbola

As illustrated in Figure 7-9, a ruler is rotatably attached at one end to the left focal point F_1. In addition, point B is marked with the blue distance $2a$ from the focal point F_1 along the edge of the ruler. A string is attached to the other end of the ruler at point A and at the right focal point F_2. The length of the string corresponds exactly to the distance from point A to point B. With the help of a pin, the thread is now stretched so that it lies against the edge of the ruler. By rotating the ruler around the focal point F_1, the pin sweeps over a hyperbola.

The Parabola—Defined as a Geometric Location

The parabola is the geometric location of all points P in a plane whose distance to a fixed point F is the same as the distance to a given directrix. As before, point F is the focal point. Figure 7-10 shows that the purple line from point P to the focal point F is equal to the purple line from point P to the green directrix. Again, it is possible to construct the parabola using string, ruler, and pencil.

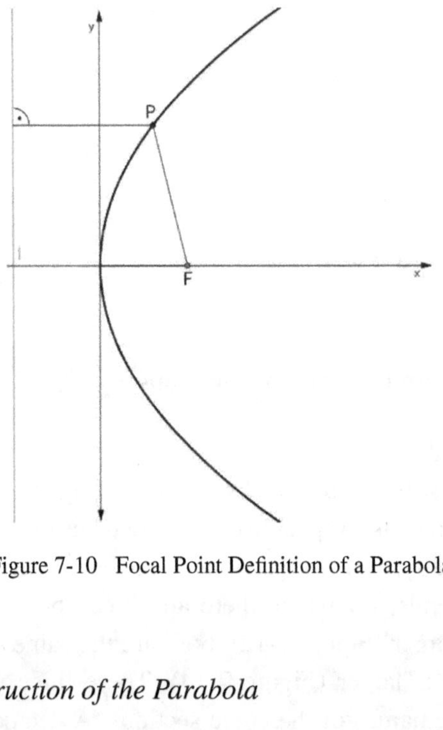

Figure 7-10 Focal Point Definition of a Parabola

The String Construction of the Parabola

A parabola—just as a hyperbola and an ellipse—can easily be created using a simple geometric construction method. To do so, an angle and a string are needed, as shown in Figure 7-11. The length of the brown string is determined by the distance between the points A and B. One end of the string is fixed at point A, and the other is fixed at the focal point F. The angle is positioned so that the angle leg can slide along the green directrix. The string is tensioned with a pin at point P in a way that it rests against the edge of the angle. By moving the angle along the directrix, the pin at point P sweeps a parabolic arc.

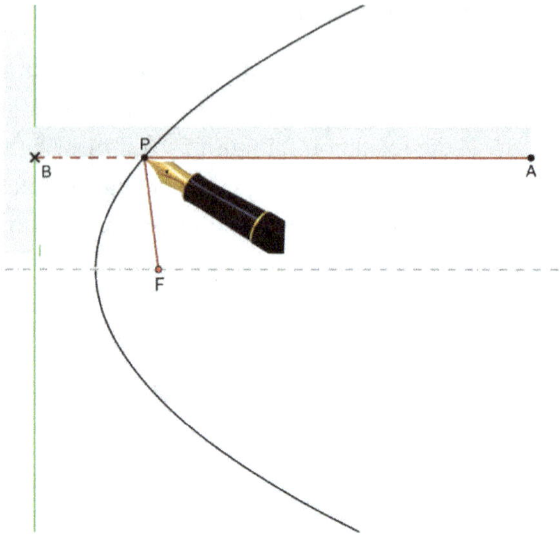

Figure 7-11 String Construction of a Parabola

Excursus: The Naming of Conic Sections

Once again, Apollonius deserves credit as he gave the conic sections their names, which are still used today. His understanding of conic sections' properties enabled him to use a specific characteristic as the basis for the respective name. Originally, the conic sections owed their names to the comparison of two quadrilaterals, for which there are three options: Either the green quadrilateral is larger (Figure 7-12), both are the same size (Figure 7-13), or purple square is larger (Figure 7-14). These three possible outcomes result in the three names of the conic sections. A detailed description can be found below.

Let us first look at the vertical line from point $F_{p'}$ to point F_p in the respective figures. This line runs through the focal point F, which is located exactly in the middle of line $F_{p'}F_p$. The partial distance from point F to point F_p (equal to the distance from F to point $F_{p'}$) indicates half the opening width of the conic sections at the height of the focal point and is described by the parameter p. As a result, the distance from point $F_{p'}$ to point F_p has a length of $2p$ and is—in mathematical jargon—referred to as blocking. This term is needed for the green rectangular area.

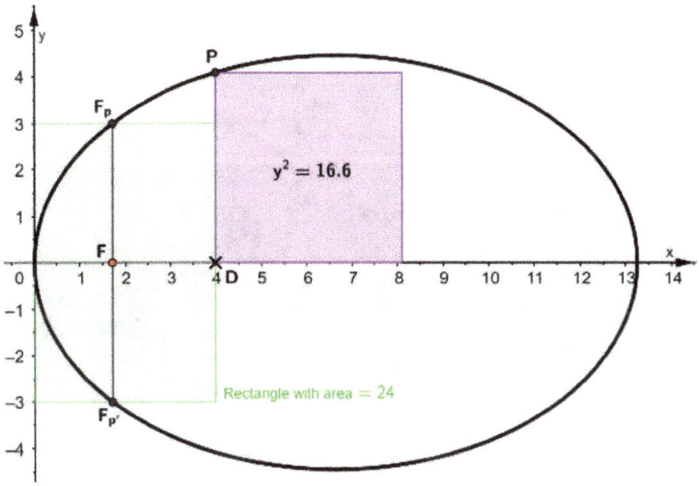

Figure 7-12 Ellipse

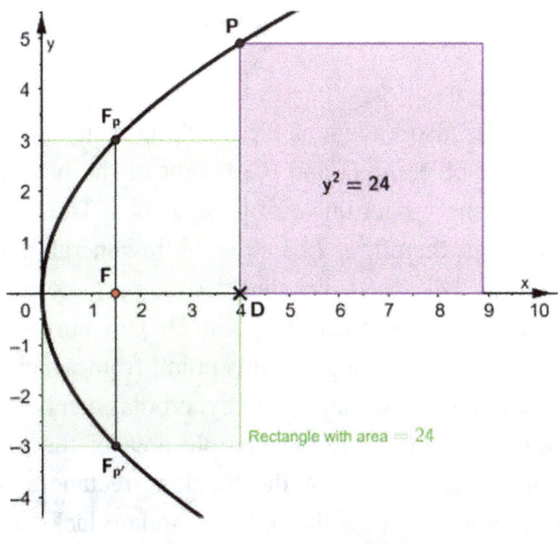

Figure 7-13 Parabola

But first, let's take a look at point D, which has the same coordinate in all three figures and is fixed on the axis. Point P is located at the same x-coordinate as point D on the conic section curve with the ordinate[4] y.

[4]The ordinate is the y-coordinate of a point.

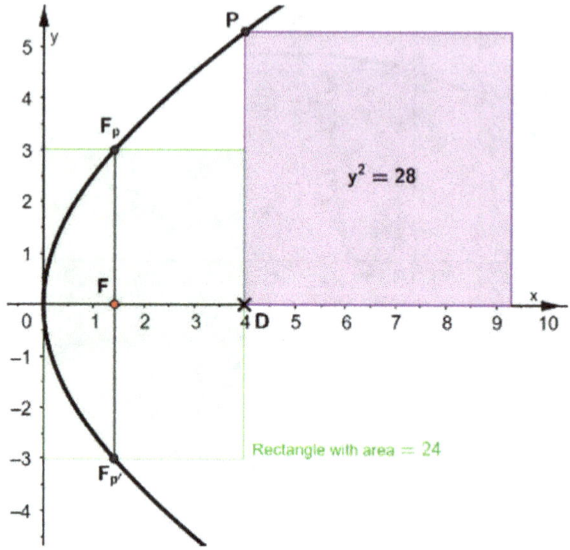

Figure 7-14 Hyperbola

The green rectangle, also known as the blocking-rectangle, has the width of the x-coordinate of point D and the height of the blocking $2p$. In all figures, $p = 3$ and the x-coordinate of point D is 4. The area of the green blocking-rectangle is, therefore, $2 \cdot 3 \cdot 4 = 24$. In general, the area formula can be expressed as $2px$. Next, we consider the purple ordinate square y^2 at the location of the x-coordinate of point D. This purple square is now compared with the green rectangle at this point, from which the following relationships for ellipses, parabolas, and hyperbolas can be concluded.

In the ellipse in Figure 7-12, the purple area of the ordinate square is smaller than the green area of the blocking-rectangle, which means $y^2 < 2px$ in general terms. As the ordinate square lacks area, this conic section curve was given the name ellipse, which derives from the Greek ἔλλειψις and means "to stand back" or "to leave out." In the parabola in Figure 7-13, the purple area of the ordinate square corresponds exactly to the green area of the blocking-rectangle. This can be formally expressed by the equation $y^2 = 2px$. The term parabola also comes from the Greek (παραβάλλειν) and means "to equal."

Finally, in the hyperbola in Figure 7-14, the purple area of the ordinate square is larger than the green area of the blocking-rectangle. Therefore, the mathematical inequality is $y^2 > 2px$. The ordinate square has an excess of area compared to the rectangle. This is why it was named hyperbola (Greek ὑπερβολή), which means "to exceed" or "to surpass."

After defining intersection curves using important terms such as focal points and directrix as well as revealing the secret of their names, we now move into the early 19th century and devote ourselves to a very special means of proof in the context of conic sections: Dandelin's spheres.

Dandelin's Spheres

Without a doubt, Dandelin's spheres are an extraordinary and ingenious proving tool in geometry. Germinal Pierre Dandelin (1794–1847) taught at the University of Liège in Belgium. He had the idea of placing one or two spheres in a straight double cone in such a way that they touch both the sectional plane at one point but also the cone's lateral surface. The resulting sectional figure corresponds to a conic section.

As an introduction, consider Figure 7-15. If the concepts presented seem comprehensible, you will be well equipped to understand the remaining

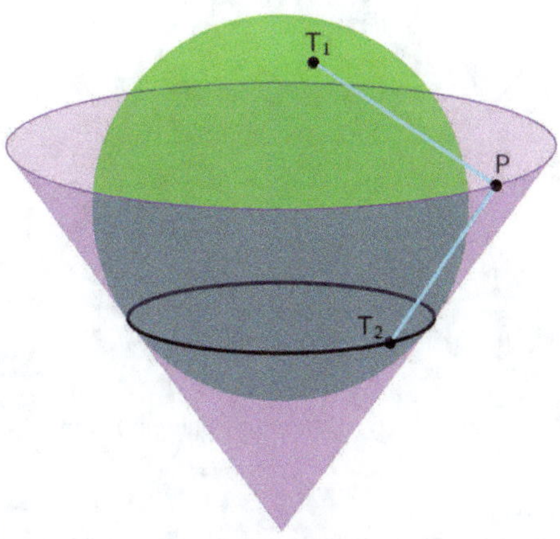

Figure 7-15

evidence, which shows that parabolas, hyperbolas, and ellipses do indeed arise when the lateral surface of a double cone is intersected.

Tangency:

In Figure 7-15, if tangents[5] are applied to a sphere from a given point, the tangent sections are of the same length. This means that the distance from point P to point T_1 is the same as the distance from point P to point T_2. Furthermore, a sphere placed in the cone touches it in the form of a circle whose plane is perpendicular to the cone's axis.

Ellipse

Figure 7-16 illustrates a cone, a sectional plane, and two spheres that are placed inside the cone. These are chosen in such a way that each of them touches the sectional plane at the red point. Now we consider the two green circles in which the two spheres touch the cone. They have a constant distance, marked by a yellow line on the cone's surface. This line intersects

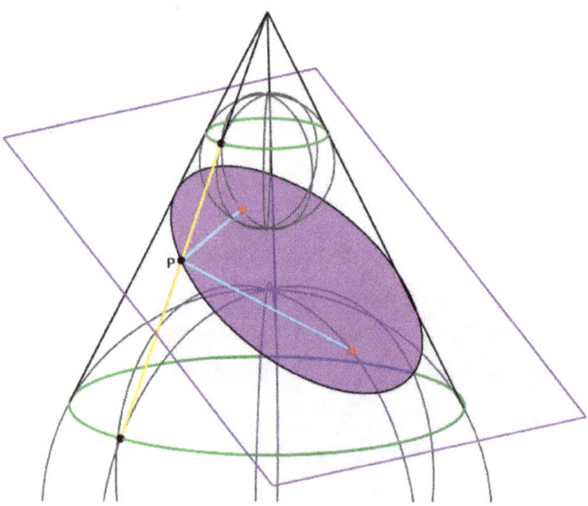

Figure 7-16 Dandelin Spheres: Ellipse

[5]A tangent is a straight line that touches a given curve or geometric object at a specific point.

the plane at a point on the intersection curve, labeled P. From this point P, two blue lines go to the two red points of contact.

Now it is time to use the first mnemonic. From point P, the distance to the red point of contact equals the distance from the generatrixes to the same sphere's circle of contact. Now let's imagine that the yellow lateral line rotates along the two circles and point P moves along the intersection curve. The sum of the distances between the two red points of contact and point P is always the same, i.e., equal to the yellow lateral line. As a result, the two points of contact can be regarded as focal points. Accordingly, the requirements of the focal point definition of an ellipse are fulfilled, and the intersection curve corresponds to the desired shape of an ellipse.

Parabola

Figure 7-17 illustrates a cone, two sectional planes, and a sphere placed inside the cone. The sphere is again selected so that it touches the plane at the red point. The first sectional plane is aligned parallel to the generatrix. The second plane is put in the position in which it passes through the green circle of contact of the sphere and the cone. It also intersects the first plane

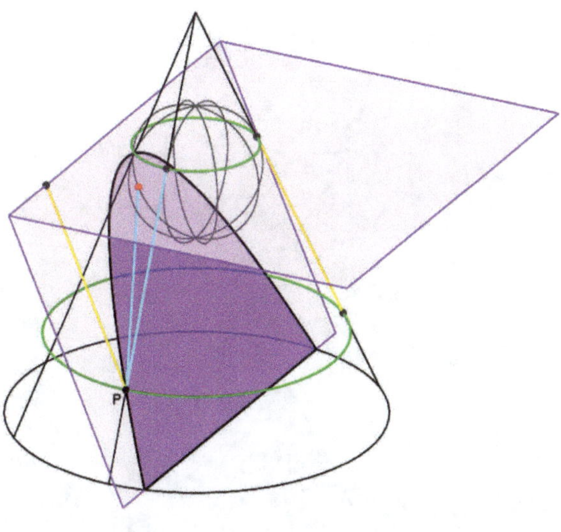

Figure 7-17 Dandelin Spheres: Parabola

in what is called a directrix. Again, a point P lies on the intersection curve. Another green circular line—parallel to the circle of contact of the sphere—runs through point P. In addition, a blue line leads from this point to the red point of contact, while the other blue circumferential line runs to the circle of contact. According to the mnemonic, the two distances are tangential sections of the same sphere, and hence of equal length. A substitute for the lateral line is found by rotating it along the two green circles. You can now see that the blue and yellow sections are exactly the same length. The yellow section is transferred to the first section plane by parallel displacement. As a result, the distance from point P to the straight line equals the distance from point P to the red focal point. The requirements of definition a parabola (see definition of the parabola) are then fulfilled and the intersection curve is the shape of a parabola.

Hyperbola

Figure 7-18 illustrates a double cone, a cutting plane, and two spheres, each of which is placed inside the cone. In contrast to the ellipse and parabola,

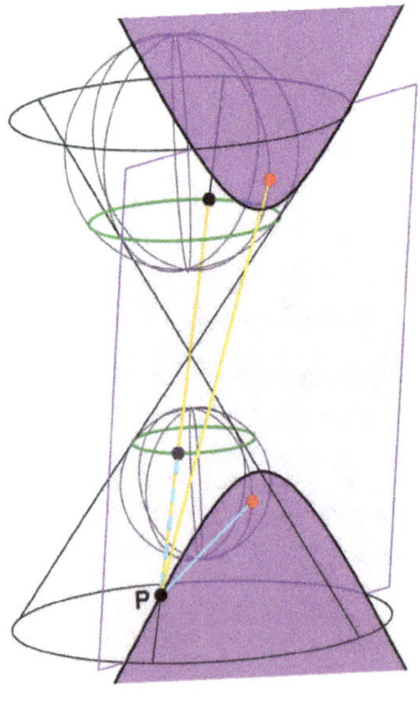

Figure 7-18 Dandelin spheres: Hyperbola

a hyperbola is created when the section plane intersects both halves of the double cone. The spheres are again chosen in order that each of them touches the sectional plane at the red point. We let P be a point on the intersection curve that is located on both the lower branch of the hyperbola and the generatrix of the double cone. From this point, a blue and a yellow line run to the two bold points of contact. According to the mnemonic, the blue distance from point P to the bold point of contact equals the blue lateral distance to the lower contact circle of the same sphere. In the same way, the yellow lateral distance from point P to the upper contact circle must be of the same length as the yellow distance from P to the upper bold contact point F. The bold points of contact are again referred to as focal points. The reasoning is the same as for the ellipse, except that it is now based on the difference in distance instead of the sum of the distances. According to the hyperbola definition, the intersection curve is ultimately a hyperbola.

Now that numerous theoretical foundations have been laid and it has been shown that an intersection of a double cone produces an ellipse, a parabola, or a hyperbola, we can explore the versatile applications of conic sections in everyday life.

Conic Sections in Everyday Life

The German writer Johann Wolfgang von Goethe (1749–1831) once said "You only see what you know." Without any doubt, these are not just empty words, but describe the truth that we only notice things about which we have background knowledge. This also applies to conic sections, as we are only aware of the intersection curves when we are familiar with their existence. Let us now focus on the conic section curves created by humankind that are a necessity in physical contexts and used in technical applications.

Shadow Play

From geometrical optics, established by Euclid, it is known that light phenomena can be described and explained in many practical cases with the help of elementary geometric principles. Geometric optics has greatly expanded

Figure 7-19 Shadow Play: Tealight
Source: With permission of Christoph Vierthaler.

our understanding of light and its behavior and helps to explain and predict light phenomena in the real world. Nowadays, tea lights and candles mainly serve the function of creating a cozy and comfortable atmosphere. For decoration and protection against wind, the light sources are often placed in glasses or other containers. If located near a wall, a well-formed shadow is created, as shown in Figure 7-19. In addition, electric lights attached to the outside of the house wall can also create captivating shadow structures, as is shown in Figure 7-20.

But how does this phenomenon arise, and which curve separates light from shadow? If a flat surface is held directly over a light source, a circle is created. The size of the circle increases proportionally to the distance between the surface and the light source. If the surface is tilted, the circle is transformed into an ellipse. Moving the surface further away and not changing the tilt angle results in a larger elliptical arc. The ratio between the major and minor axes remains constant. By increasing the inclination of the surface, the focal points of the ellipse move apart and make the shape of the ellipse longer. This effect intensifies until one focal point finally moves

Figure 7-20 Shadow Play: Electric Lights
Source: With permission of Christoph Vierthaler.

to infinity and the surface is aligned parallel to one of the rays emerging from the edge of the container or lampshade. This phenomenon finally leads to a parabola. However, when looking at the shadow structure on the wall, it is often neither an ellipse nor a parabola but the branch of a hyperbola. This is because, with the light source being placed at the center of a container or glass, the nearby vertical surface (e.g., a wall) cuts a hyperbola out of the light cone.

One can easily try out the play of light and shadow by holding a sheet of paper over a flashlight. Tilting the paper creates various conic section lines in the form of an ellipse, parabola, or hyperbola.

Architectural Masterpieces

Buildings or sculptures in the shape of a hyperbola play an important role in architecture for a variety of reasons. One example would be the Metropolitana Nossa Senhora Aparecida cathedral in Brasília, Brazil, shown in Figure 7-21. Built between 1958 and 1970 and dedicated to the Blessed Virgin Mary, the name refers to Aparecida, the largest place of pilgrimage

Figure 7-21 Cathedral Metropolitana Nossa Senhora Aparecida
Source: Rodrigo de Almeida Marfan. CC-BY-SA 4.0 (https://creativecommons.org/licenses/by-sa/4.0/deed.en).

in Brasília. The Brazilian architect Oscar Niemeyer (1907–2012) shaped the cathedral in the form of a single-shell hyperboloid,[6] whose external appearance seems to symbolize praying hands. The 16 concrete columns arranged in a row are not only aesthetically pleasing, but also offer structural stability and technical advantages. The unique shape makes it possible to create large spaces without supporting columns. In addition, the circular base extends to a diameter of 70 meters and can accommodate over 4000 people.

Another impressive masterpiece is the sculpture *Mae West* (Figure 7-22), which can be found at the Effnerplatz in Munich-Bogenhausen, Germany. The artist Rita McBride (1960–) named the sculpture after the American film actress Mae West. Made of plastic, the bar frame reaches a height of 52 meters and also takes the form of a single-shell hyperboloid. To prevent the formation of icicles in winter, electric heating has been installed in the individual tubes.

[6]This mathematical solid is created when a hyperbola rotates in space around its vertical axis. A single-shell hyperboloid is therefore the 3-dimensional equivalent of a hyperbola.

Figure 7-22 The Sculpture *Mae West*
Source: Martin Falbisoner.

Planetary Orbit

The infinite space of the universe hides some of the most fascinating secrets mankind has ever discovered. One of those is found in the seemingly simple orbits of the planets around the sun. What was once considered a riddle was revealed by the German astronomer Friedrich Johannes Kepler back in the 16th century: The planets orbit the sun along elliptical paths, with the sun not being in the center of the ellipse but at the so-called focal point (Kepler's first law). Moreover, Kepler's first law was an important prerequisite for Sir Isaac Newton's formulation of classical mechanics as well as the law of gravitation.

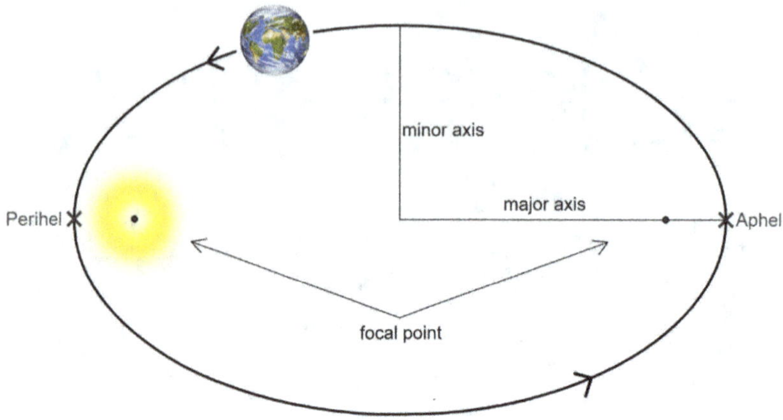

Figure 7-23 Earth's Elliptical Orbit

As can be seen in Figure 7-23, the Earth's orbit around the sun corresponds to an ellipse. However, the deviation to a circular orbit is relatively small, as the minor semi-axis is only about 40,000 kilometers shorter than the major semi-axis. One should also bear in mind that the seasons are not influenced by the distance between Earth and Sun. It is the Earth's axis, with an angle of inclination of about 23.5 degrees in relation to the orbit around the sun, that determines the seasons.

Nevertheless, the winter half-year of the Northern Hemisphere is about seven days shorter than the summer half-year. As expected, the reason for this is the elliptical orbit. At the beginning of January, the Earth is closest to the Sun (perihelion) with about 147 million kilometers distance. In July, on the other hand, it is farthest from the sun (aphelion) with a distance of about 152 million kilometers. According to Kepler's second law, the Earth moves slightly faster when it is close to the Sun. That is the reason why the winter half-year is shorter in the Northern Hemisphere.

Trajectories

Italian mathematician Galileo Galilei (1564–1642) formulated equations for throwing parabolas' trajectory that disproved previous assumptions about the trajectory of cannonballs. Before this, people believed that the trajectory of a cannonball consisted of two straight lines connected by a circular curve. Galileo also discovered that, without friction, all bodies, regardless of their mass, do not only move in the same way during a free fall but also experience the same acceleration. All over the earth's surface, this acceleration is approximately $9.8\ \frac{m}{s^2}$. In recognition of Galileo Galilei's contribution, this particular acceleration due to gravity is given the formula letter g. In addition, Galileo proved that the trajectory of a thrown object in a vacuum describes an exact parabola. In the following section, two forms of trajectories, the horizontal throw and the oblique throw (which represents the general case) are discussed. While the mathematical equations of both parabolas are explained in detail, physical examples of everyday life emphasize their practical relevance.

Horizontal Throw

We can consider how an item behaves when it is thrown parallel to the horizon. Two forces play a decisive role. One is the gravitational force of the earth, which exerts a uniform acceleration downwards in the y-direction. This means that the thrown item moves with a constant acceleration in the direction of the earth's surface. On the other hand, the throwing force acts to give the body the necessary uniform straight-line movement in the horizontal x-direction. This means that the item does not experience any changes in its horizontal speed and is, therefore, propelled in this direction at a constant speed. The resulting trajectory is a parabola with a launch angle of $\alpha = 0°$. In the following, the horizontal throw is described mathematically using an equation. The procedure is based on the following methodology.

We use the equation of uniform motion for the movement in x-direction, while the equation of free fall is required for the movement in y-direction. The latter is caused by the gravitational pull of the body vertically downwards with the acceleration due to gravity. Finally, the two equations are linked.

For uniform straight-line movement in the x-direction, $s_x = v_0 t$, where s_x describes the throw distance as a function of time t and v_0 describes the constant initial velocity. If the equation is rearranged according to the variable t, then $t = \frac{s_x}{v_0}$ is obtained.

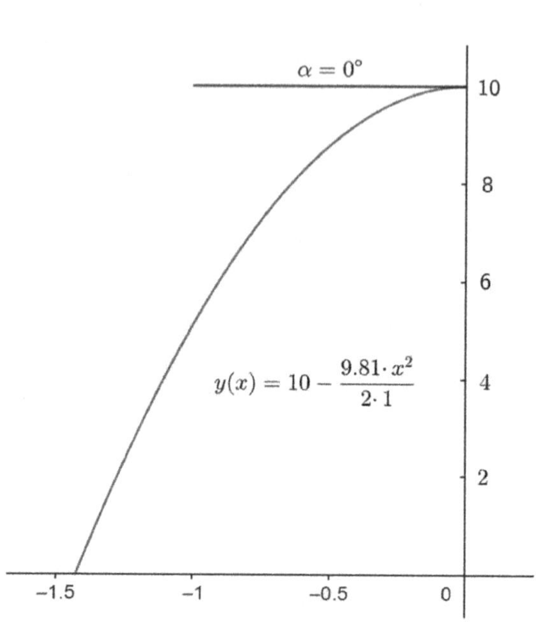

Figure 7-24 Horizontal Throwing Parabola

The formula for uniformly accelerated movement in the y-direction is $s_y = h_0 - \frac{1}{2}gt^2$, where s_y describes the height of the throw at time t, and h_0 the height of the drop. The expression $\frac{1}{2}gt^2$ describes the free fall with t as the fall time. The variable g represents the acceleration due to gravity.

By inserting the first equation—which has previously been solved for t—into the second equation for the uniformly accelerated movement, we obtain a downward-opening parabola $y(x) = h_0 - \frac{1}{2}g(\frac{s_x}{v_0})^2$, as shown in Figure 7-24. In Figure 7-24, the drop height h_0 is 10 meters, and the initial velocity v_0 is $1\frac{m}{s}$. Typical examples of horizontal throws include a ball that falls off a table or the stream of water from a horizontally directed spout, such as in the garden of the Belvedere Palace in Vienna, shown in Figure 7-25.

A tricky riddle in the context of horizontal throwing: Is it true that a ball thrown straight ahead takes longer to reach the ground than one dropped vertically from the same height? The answer can be found later in this chapter.

Figure 7-25 The Shape of a Throwing Parabola
Source: Zátonyi Sándor. CC-BY-SA 3.0 (https://creativecommons.org/licenses/by-sa/3.0/deed.en).

Oblique Throw

The oblique throw now represents the general case. Again, the entire motion is composed of a uniform motion in the x-direction and a uniformly accelerated motion in the y-direction. In contrast to the horizontal throw, the body is now thrown upwards in an oblique direction with a certain angle of release α; due to gravity it is slowed down and then falls down again. In the following, the oblique throw is described by means of a mathematical equation.

The uniform straight-line movement in x-direction can be explained almost analogously to the horizontal throw. However, the specific launch angle α is now added, which is why the equation can be expressed using the cosine function: $s_x = v_0 t\cos(\alpha)$.[7] Here, s_x describes the throwing distance as a function of the time t, while v_0 represents the velocity of the throw. In terms of the variable t, the expression is $t = \frac{s_x}{v_0 \cdot \cos(\alpha)}$.

The formula for the uniformly accelerated movement in the y-direction is slightly more complex. The equation $s_y = h_0 + v_0 \sin(\alpha)t - \frac{1}{2}gt^2$ applies to the throw height as a function of time t. The variable h_0 describes the height of the drop, the expression $v_0\sin(\alpha)t$ corresponds to the throw upwards, and $-\frac{1}{2}gt^2$ again explains the free fall downwards.

As with the horizontal throw, we substitute for t in the second equation to get the trajectory of a parabola with $y(x) = h_0 - \frac{1}{2}g\frac{x^2}{(v_0 \cdot \cos(\alpha))^2} + \tan(\alpha)x$, as can be seen in Figure 7-26. For this example, a launch height h_0 of 2 meters,

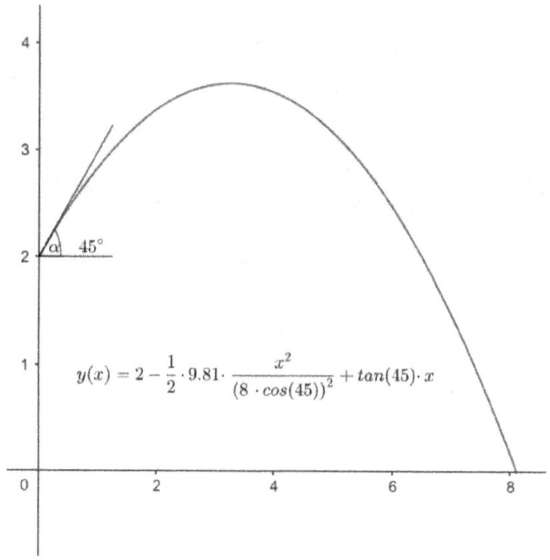

Figure 7-26 Oblique Throwing Parabola

[7]The item is thrown at a certain angle α to the x-axis with an initial velocity v_0. This initial velocity can be decomposed into x-direction and y-direction. Let us now consider a right-angled triangle with the hypotenuse v_0, the two cathets $v_{0.x}$ and $v_{0.y}$, and the angle α. By rearranging the equation $\cos(\alpha) = \frac{adjacent\ side}{hypotenuse} = \frac{v_{0.x}}{v_0}$, we obtain the initial velocity in the x-direction as $v_{0.x} = \cos(\alpha) \cdot v_0$. The velocity at a certain point in time t is therefore $v_x(t) = v_0 \cdot \cos(\alpha)$. Since we already know that a uniform movement takes place in the x-direction, we can formulate the equation distance = velocity · time. This results in $s_x = v_0 \cdot t \cdot \cos(\alpha)$.

Figure 7-27 The Shape of a Throwing Parabola
Source: GuidoB. CC-BY-SA 3.1 (https://creativecommons.org/licenses/by-sa/3.0/deed.en).

a launch velocity v_0 of $8\frac{m}{s}$, and a launch angle of 45° were chosen. Typical examples of the oblique throw include a ball's impact in shot put and a spout of water from a fountain, as can be seen in Figure 7-27.

The answer to the riddle: It is not correct. Let us first consider the ball that is dropped vertically from the same height. Due to gravity, it moves vertically downwards towards the ground. Accordingly, the previously mentioned equation $s_y = h_0 - \frac{1}{2}gt^2$ applies to the y-coordinate. In contrast, the x-coordinate remains constant as there are no horizontal forces acting on the ball.

Let us look at the other ball, which is always thrown not only from the same height but also straight ahead. The special thing about the horizontal throw is that we can consider the two components as being completely independent of one other. For the y-component, the equation is the same as for the ball that was dropped vertically. For the x-component, the equation is the same as for the uniform motion. The solution to the puzzle lies in the fact that we are only interested in the y-component, that is, the falling time. Depending solely on the initial height h_0 and the gravitational force g, the falling time is the same for both throws. As a result, the two balls reach the ground at the same time.

Why Bridges Like to Be Parabolas

Bridges span rivers, valleys, or even straits. Considering their span, many of these constructions are based the parabola. In many cases, the parabolic structure is revealed in the supporting arch or in the suspension cable. We shall consider the latter.

Let's start with the principle on which the suspension bridge is based. In Figure 7-28, the deck is suspended from the robust, parabolic suspension cables using numerous vertical cables. These suspension cables are in turn stretched over two towers, which support the weight of the bridge by means of the lateral anchors at the ends of the bridge. This construction converts the purely vertical weight of the bridge into a vertical component and a horizontal component in the suspension cable. This converting process takes place at the towers, at the anchorages, and at every other cable point.

When constructing a bridge, mathematical analysis must answer a series of important questions: How high should the towers be, and what load will the towers have to carry? How high should the bridge be built above the water so that shipping traffic is not obstructed? It is also particularly important to investigate the horizontal and vertical forces to which the suspension cables are exposed. We will now focus on the latter.

You might think that the suspension cables are subjected to the greatest tensile force[8] at the lowest point, which is exactly in the middle of the two towers. However, this is not the case since the horizontal component remains constant throughout the suspension cable due to the equilibrium

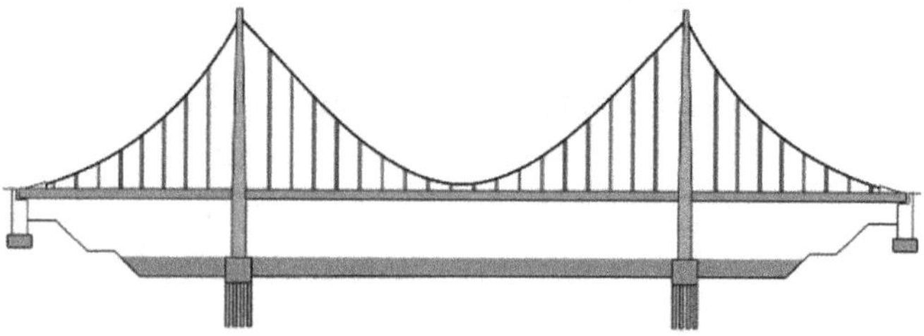

Figure 7-28 Principle of a Suspension Bridge

[8]Tensile force is the force that pulls on a body.

conditions.[9] In contrast, the vertical component varies depending on the distance from the middle of the suspension cable to a fixed point on the suspension cables. At the lowest point of the suspension cable, the vertical component is 0, as the distance to a point on the suspension cable is also 0. At other points, however, the component is always greater than 0. The vertical component then grows with increasing distance from the center. This means that it reaches its maximum at the edges near the towers. In other words, the bridge pulls down with its entire weight at the edges. At other points, however, only the part of the bridge that lies between the center of the suspension cable and this point pulls down.

The amount of slack in the suspension rope also plays a decisive role. Generally speaking, the greater the sag, the lower the forces in the suspension rope. This applies in particular to the constant horizontal component in the rope. Let us imagine two pulleys at the upper end of the two towers through which the suspension rope runs. The more the suspension rope is "tensioned," the more outward force is naturally required.

The first suspension bridges were built in the 19th century. One of the oldest still in use for road traffic is the Union Bridge in England (see Figure 7-29), which was completed in 1820. With a span of around 130 meters, it is quite a historical masterpiece. A milestone in the development of suspension bridges was the well-known George Washington Bridge over the Hudson River in New York, which was completed in 1931 (see Figure 7-30). With an enormous span of 1076 meters, it surpassed all previously constructed suspension bridges in terms of size. This construction laid the foundation for new standards, as was the case with the famous Golden Gate Bridge in San Francisco. The bridge spans about 1280 meters, has a total length of 2737 meters, and was completed in 1937.

As of 2024, the world's longest suspension bridge is currently located in the Sea of Marmaraand spans the Dardanelles in Turkey. Completed in 2022, the 1915 Canakkale Köprüsü Bridge has an impressive total length of 3869 meters and a span of more than 2020 meters. The construction only took five years (see Figure 7-31).

[9] This means that the horizontal forces must be in equilibrium at every point of the bridge and thus cancel each other out.

Figure 7-29 Union Bridge in England
Source: DeFacto. CC-BY-SA 4.0 (https://creativecommons.org/licenses/by-sa/4.0/deed.en).

Figure 7-30 George Washington Bridge over the Hudson River
Source: Jim Harper. CC-BY-SA 2.5 (https://creativecommons.org/licenses/by-sa/2.5/deed.en).

Figure 7-31 The 1915 Canakkale Köprüsü, the World's Longest Suspension Bridge
Source: Zafer. CC-BY-SA 4.0 (https://creativecommons.org/licenses/by-sa/4.0/deed.en).

Excursus: Reflection Property of the Conic Sections

For all conic sections, the rays originating from a focal point are reflected at the curve in a special way. A ray that has its origin at focal point F_1 is reflected at point P on the conic section curve in such a way that it runs along the direction of the straight line from point F_2 to point P. As already known, F_2 is the second focal point. In order to gain a deeper understanding, it is helpful to visualize the reflection property. Figure 7-32 shows an ellipse. The ray emanating from focal point F_1 is reflected at point P on the ellipse curve in such a way that it passes through the other focal point F_2. It is then again reflected on the curve and returns to the focal point F_1.

Figure 7-33 shows a hyperbola for which the ray emanating from the focal point F_1 is reflected at point P on the hyperbolic curve as if it came from the other focal point F_2.

Finally, the parabola in Figure 7-34 shows a remarkable reflection. Starting from the focal point F, the ray is reflected at point P on the parabola curve so that it continues to run parallel to the axis. The reason for this is

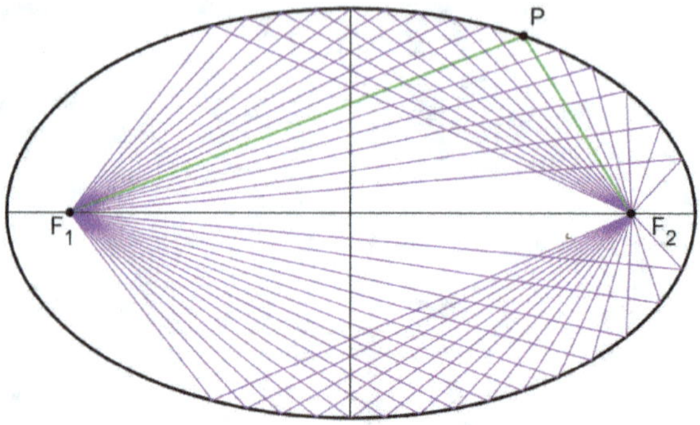

Figure 7-32 Reflection Ellipse

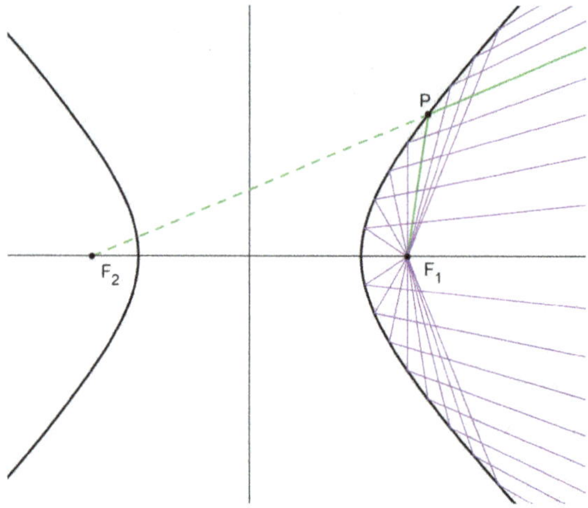

Figure 7-33 Reflection Hyperbola

that the second focal point is at infinity. It should be noted that this reflection works in both directions. This means that axis-parallel incident rays are reflected at the parabolic curve and collected at the focal point F.

In everyday life, we come across a variety of shapes and structures, often without realizing that they are based on mathematical principles. The aim of this section is to present fascinating examples of conic sections' reflection property. While we encounter some of these examples only rarely or not at all, others are deeply rooted in our everyday lives and accompany us regularly.

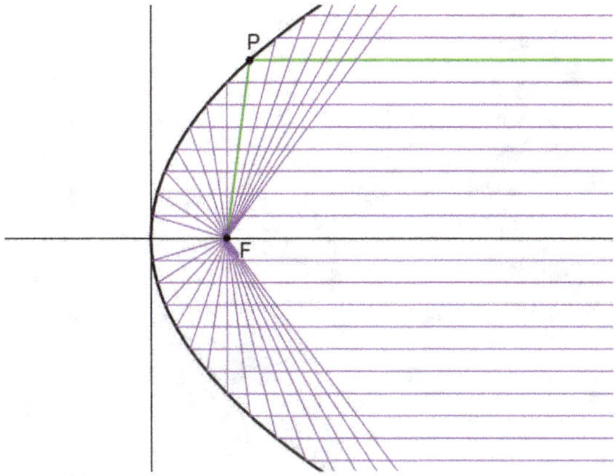

Figure 7-34 Reflection Parabola

Ellipse

Whispering Galleries

In the field of architecture and acoustics, a remarkable connection between form and sound can be noticed in several historical buildings. A striking example is Grand Central Terminal in Manhattan, New York. The terminal was built in 1913 and is the largest train station in the world. It also needs to be mentioned that the train station is used by around 750,000 visitors daily. In the lower part of Grand Central there are four square, vaulted archways, as shown in Figure 7-35. If you whisper into one of these, it is possible for the person standing in the diagonally opposite archway to hear clearly what is being said, despite a distance of more than nine meters. The reason is that the sound waves of the person speaking are reflected by the elliptical building ceiling in such a way that they gather again at the second focal point, i.e., the listener. Figure 7-36 illustrates how the German mathematician Athanasius Kircher (1602–1680) recognized that the geometric shape of rooms influences acoustic behavior.

The New York train station may present a hectic and noisy atmosphere; yet a quiet, silent secret lies within. Whispering galleries also reveal themselves in other places around the world, such as the Mormon Tabernacle in Utah, the Gol Gumbaz Mausoleum in India, Saint Paul's Cathedral in London, and the National Statuary Hall of the United States Capitol.

Figure 7-35 Elliptical Space – Grand Central Terminal
Source: Michael Freeman.

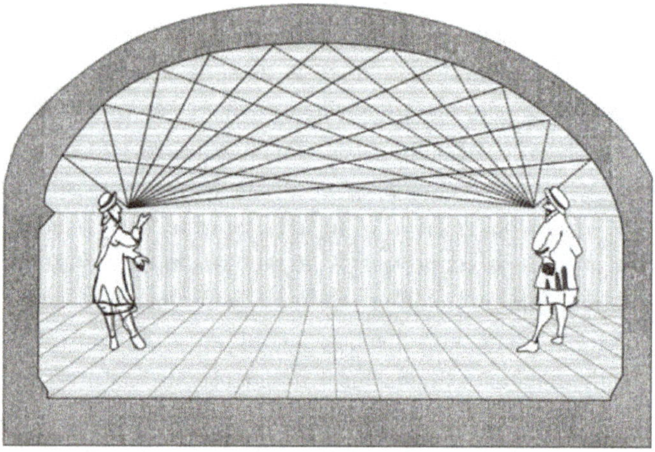

Figure 7-36 Elliptical Space—Heidelberg's Echo
Source: Trever Cox, with permission.

Lithotripsy: Kidney Stone Crusher

The most modern application of this principle of reflective properties is in the kidney stone crusher. The elliptical mirror contains two electrodes that generate shock waves at one focal point. These meet again at the other focal point, where the kidney stone is located, and trigger a voltage at the stone. The high energy at the focus of this shock wave shatters existing kidney, saliva, or gallstones, allowing the fragments to leave the body naturally. The advantage of this method is that these stones can be removed completely without surgery, without anesthesia, and (almost) without any pain. This remarkable method was first successfully carried out in the 1980s by physicians at the University Hospital Grosshadern in Munich, Germany. As can be seen in Figure 7-37, the first devices had a tub filled with water in which the patient was placed. Since 2005, the tub is no longer needed, and the devices look similar to modern X-ray machines.

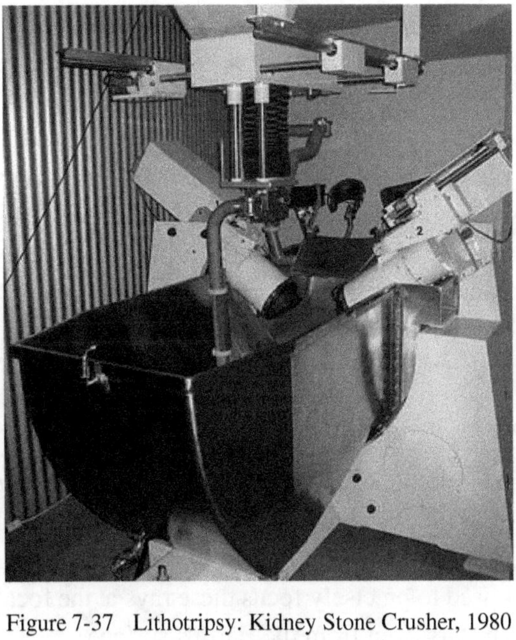

Figure 7-37 Lithotripsy: Kidney Stone Crusher, 1980

Compact Blast Furnaces

A mathematical feature is also used in compact blast furnaces. The heating element is located at the focal point of the ellipse, while the sample to be heated is located at the other focal point. The heat radiation emitted by the heating element is reflected back to the sample at the wall, which has the shape of an ellipsoid.[10] A major advantage is the uniform heating of the sample, as it is irradiated with heat from all sides. A furnace like that was used in American space shuttles.

Parabola

Parabolic Mirror

Parabolic mirrors are applied in various ways. Examples are its use as a car headlight, as a parabolic trough in a solar power plant, and its function as an essential component of telescopes. Due to its ability to bundle beams in highest precision, the parabolic reflector is often used as a focusing optic.

Car Headlights

The headlight on a passenger car often takes the form of a parabolic mirror. For this purpose, a point-shaped light source is positioned exactly at the focal point of the parabolic mirror. The beams of the bulb are reflected parallel to the axis in the parabolic housing and thus form the headlight beam (see Figure 7-38). This enables the road in front of the car to be illuminated.

Radio Telescope

In contrast to the spotlight emitting light, the telescope bundles the light rays. Especially when observing distant objects, such as stars in space, the light rays run almost parallel into the structure. A parabolic mirror inside the telescope is used to precisely focus these rays at the focal point, making it possible to see stars even from the Earth's surface.

[10] An ellipsoid is the 3-dimensional equivalent of an ellipse.

Figure 7-38 Car headlights: Daimler
Source: Christoph Vierthaler, with permission.

Figure 7-39 shows Parkes Observatory, located north of the City of Parkes in Australia. The telescope, with a parabolic mirror of over 60 meters, was put into operation in 1961 and provided the transmission of the television images of the Apollo 11 moon landing in 1969. The incoming beams were collected in the receiving antenna, which, in turn, is located at the focal point. Then, the signal was sent out for the region around it. This technique is also applied to satellites, as buildings are equipped with the typical parabolic antennas known as satellite dishes.

Andasol Solar Power Plant

Parabolic mirrors can be used to concentrate parallel incident sunlight in parabolic cylindrical troughs. The collectors consist of curved mirrors that concentrate the sunlight onto an absorber tube running through the focal point with a remarkable accuracy of over 94 percent. In these tubes, the concentrated radiation is converted into heat, which is transferred to oil circulating through the tubes. The oil, reaching temperatures of up to 400 degrees

Figure 7-39 Parkes Observatory
Source: Stephen West. CC-BY-SA 3.0 (https://creativecommons.org/licenses/by-sa/3.0/deed.en).

Celsius, generates steam via heat transfer, which is used to drive the power plant turbines.

Used in the Spanish province of Granada, this technology is not only the first parabolic trough power plant in Europe, but also the largest of its kind in the world. The complex consists of three power plants: Andasol 1, Andasol 2, and the most recent solar power plant, Andasol 3, which has been in operation since September 2011. Each of the three solar fields is equipped with a total of 209,664 parabolic mirrors and covers an area of approximately 210 soccer fields. At full capacity, the power plant supplies electricity for 600,000 people.

Index

www.ingramcontent.com/pod-product-compliance
Lightning Source LLC
LaVergne TN
LVHW020234210926
848481LV00016B/1947